Madness

Madness

A Philosophical Exploration

JUSTIN GARSON

OXFORD
UNIVERSITY PRESS

OXFORD
UNIVERSITY PRESS

Oxford University Press is a department of the University of Oxford. It furthers
the University's objective of excellence in research, scholarship, and education
by publishing worldwide. Oxford is a registered trade mark of Oxford University
Press in the UK and certain other countries.

Published in the United States of America by Oxford University Press
198 Madison Avenue, New York, NY 10016, United States of America.

© Oxford University Press 2022

First published in paperback 2024

Library of Congress Control Number: 2022930038
ISBN 978–0–19–761383–2 (hardback)
ISBN 978–0–19–778131–9 (paperback)

DOI: 10.1093/oso/9780197613832.001.0001

Paperback printed by Marquis Book Printing, Canada

For Noah, whom I love dearly:
May He give you the desire of your heart
And make all your plans succeed

What, then, should be the beginning (of our speech), or, just as when we drink from the well, raising the cup to the lips we never stop again, but pour in the liquid all at once, so too should our speech say everything all at once?

—Aristides, Oration XXXIX

Contents

Acknowledgments

For valuable feedback on some of the material in this book, I'm indebted to Mel Andrews, Brandon Conley, Andrew Evans, Luc Faucher, Ginger Hoffman, Scott Keller, Anya Plutynski, Sandy Shapshay, Katie Tabb, and two anonymous referees. I've benefited enormously from comments and suggestions from audience members at meetings of the Association for the Advancement of Philosophy and Psychiatry, the History of Science Society, the Peter Sowerby Foundation for Philosophy and Medicine, the Philosophy of Psychiatry Webinar hosted by the University of Québec, and the Philosophy in the Library series at Brooklyn Public Library.

I'm also grateful to Lucy Randall at OUP for her guidance on this project, and to Hannah Doyle for her editorial support. I also wish to thank Benoit Genest Rouillier for permission to use his artwork for the book cover, and Rachel Perkins at OUP for the cover design.

It would have been impossible to write the manuscript without the institutional support of Hunter College, which granted a fellowship leave during the 2019–2020 academic year for that purpose. I also want to thank Kelly Yu, who provided valuable research assistance in the early stages of the project. Kelly was supported by Hunter College through an Undergraduate Research Fellowship.

Most of all, I'm thankful to my wife, Rita, for her conversations, her kindness, and her patience during the writing process. I'm also grateful every day for the privilege of watching my two sons, Elias and Noah, grow into such extraordinary young men.

Introduction

I.

This is not an exercise in the history of science. It is not an attempt to *situate* the great thinkers of madness in a coherent and original narrative. Rather, it is a recovery mission. It is an attempt to extract from those theorists a certain vision, to home in on a style of thinking. This way of thinking is ever-present but sometimes submerged or buried so deeply that it is difficult to recognize for what it is.

It is a way of thinking that sees in madness—either in madness in its entirety, or in some of the distinctive forms that madness takes—the working out of a hidden purpose; instead of a defect, it sees a goal-driven process, a well-oiled machine, one in which all of the components work *exactly as they ought*. Madness is not the aberration of teleology, or the failure of a function, but its satisfaction. We might be tempted to say of this way of thinking that it is the precise antithesis to "medicalization"—but for the fact that we do not yet know what medicalization is.

Let us recommend some labels. I use the label *madness-as-strategy* to designate this special way of thinking that sees, in the mad, a purpose being fulfilled, a movement toward a goal, a machine operating as it ought. This way of thinking takes as the highest ambition of anyone who purports to heal the mad, and as the very key to therapy, the ability to discern the secret purpose that madness is trying to fulfill, and, once discerned, to either assist madness in the satisfaction of that purpose, or, if absolutely necessary, to divert its course toward a more psychologically and socially beneficial outcome. Put differently, those thinkers and theorists who fall under the madness-as-strategy label are able to see that some people are truly *mad by design*, not by defect or accident.

What, then, would be the alternative way of thinking, this opposing form of talking to and about madness, of treating madness? I call it *madness-as-dysfunction*. This way of thinking holds, as axiomatic, the following proposition: when someone is mad, it is because something has gone wrong inside

Madness. Justin Garson, Oxford University Press. © Oxford University Press 2022.
DOI: 10.1093/oso/9780197613832.003.0001

of that person; something in the mind, or in the brain, is not working as it ought. Madness results from the breakdown of a well-ordered system; it is a defect or a dysfunction. It represents the failure of the system to achieve its natural end.

Kant, in his youthful "Essay on the Maladies of the Head," gave us, perhaps, the clearest expression of madness-as-dysfunction in the history of psychiatry. For Kant, the mind was segmented into three main faculties or powers: the faculties of experience, judgment, and reason, each having its characteristic function. *Madness happens when one or more of these faculties fails to perform its function.* Necessarily, for Kant, there are as many basic forms (*Hauptgattungen*) of madness as there are faculties of the mind, since each form of madness simply represents a defect in the corresponding faculty. Derangement (*Verrückung*) corresponds to a defect in the faculty of experience; dementia (*Wahnsinn*) to a defect in judgment; insanity (*Wahnwitz*) to a defect in reason.

Madness-as-dysfunction did not originate with Kant. It has been handed down to us throughout the centuries, from the early Hippocratics, who taught that different forms of madness result from the corruption or perversion of the body's various humors, to the *Diagnostic and Statistical Manual of Mental Disorders* (*DSM*), which sees, in the mad, a "behavioral, psychological, or biological dysfunction," and even to the recent Research Domain Criteria (RDoC) project, according to which, simply put, "mental disorders can be addressed as disorders of brain circuits."[1]

By the last third of the twentieth century, madness-as-dysfunction became an intellectual *vise*, a mesh that girdles our thought so tightly that it has become nearly impossible to see it *as* a tradition, as a *Denkstil*, as one among other ways of observing and interacting with the mad. *The sole purpose of this book is to drive a wedge into that vise, to allow a sliver of light to break through, to allow us to glimpse that what we have taken for so long to be the truth of madness is simply a possible mode of interpretation, one among others.* The question is *not* one of destroying madness-as-dysfunction or refusing to apply it where it deserves to be applied, but *to make it possible to even raise the question* of whether madness-as-dysfunction ought to be applied in any particular case.

Finally, this book is not a genealogy. Its overriding concern is *not* "How, historically and empirically, did we arrive at this point where we cannot see madness-as-dysfunction as a specific, historically contingent thought style?" While it does follow a roughly chronological order, it is not a narrative: *this happened, which caused this to happen, which made this happen, which explains where we are now.* Rather, its overriding concern, and its principal method, is to engage with specific theorists of madness—philosophers, physicians, theologians—and to extract from each of them a teleological core, an attempt to think of madness as a strategy for accomplishing a goal. The thinkers and theorists this book surveys are not always bound together by cause-and-effect, by chains of influence, "X, who read Y, who appropriated Z . . ." Rather, they are joined by a secret resonance, a *spiritual brotherhood*. It is this secret resonance that I seek to make known, to render transparent and to subsume under the clarity of a single concept: *madness-as-strategy*.

It is for this reason that the book is not a work of history, but an exercise in concept building. It seeks to exploit historical texts as material for constructing a new concept, rather than to embed them in a sociohistorical nexus or to present them as a "product of their times." By the end of the book, if all goes well, if the book serves its function, does its job, we will have rescued this *Denkstil*: *madness-as-strategy* will come to life; it will grow wings; it will resume its rightful place among other styles of thought. We can then use it as a way to loosen the grip of madness-as-dysfunction—not, I repeat myself, for the purpose of destroying it, but for the purpose of forcing it to make space for another way of thinking. Rather than a genealogy, or even an argument, the book is something like a collection of interlocking meditations, each of which gestures toward the same underlying principle, a hidden telos of madness.

II.

To clarify the distinction between madness-as-strategy and madness-as-dysfunction, it is useful to contrast it with another distinction with which it is easily confused: mind and body, mental and physical, psychogenic and biogenic. This latter distinction has played an important role in thinking about madness; in fact, some historians have depicted *all* of psychiatry's history as little more than a pendulum swinging back and forth between biogenic and psychogenic "perspectives."[2] This is decidedly not the project being advanced here.

Crucially, the split between these two ways of thinking, madness-as-strategy and madness-as-dysfunction, can in no way be reduced to, identified with, or subsumed under the more familiar split between biogenic and psychogenic approaches to psychiatry, that is, those approaches that take the nervous system as the starting point of research and those that take the mind, abstracted from the nervous system, as the starting point. The latter split is grounded in the difference between body and mind; the former split is grounded in the difference between teleology and dysteleology, function and dysfunction, accomplishment and miscarriage. These two distinctions, that between mind and body, and that between teleology and dysteleology, crosscut each other; they are logically independent of each other. When proponents of RDoC tell us that the various forms of madness are but *biological dysfunctions*, they are effectively superimposing two logically distinct categories, the category of the *biological* and the category of the *dysfunctional*.

But these two categories slide apart. Each can be conceived clearly and distinctly apart from the other. For example, Kant held a thoroughly psychogenic point of view on madness—he had almost nothing to say about its biological basis—yet, still, he was a proponent of madness-as-dysfunction. Madness, for Kant, involves a dysfunction in a faculty of the mind, regardless of its physical basis. The cognitive behavioral therapists of today can, like Kant, be said to take a psychogenic approach, but they also, as a rule, endorse madness-as-dysfunction, for they see in the various forms of madness *dysfunctional patterns of thinking and behavior*. Hence madness-as-dysfunction crosscuts the psychological/biological divide.

So, too, does madness-as-strategy. To take but one example, Freud's psychoanalytic system was not only psychologically oriented, but thoroughly teleological. For Freud, the core forms of madness, such as hysteria, obsessional neurosis, and delusion, represented the deviant fulfillment of an unconscious wish. In each of these forms, the "psychic apparatus" *is striving to accomplish a certain function or end*. These forms are only so many different strategies for achieving that end. Other thinkers who adopted madness-as-strategy assumed a *biological* orientation. Kurt Goldstein, for example, taught that all of the disturbances of mental life are various modalities of what he called the self-actualization of the organism (*Selbstverwirklichung des Organismus*). Compulsive behavior, for example, must be seen as the organism's attempt to modulate its environment in such a way as to minimize the ever-present threat of catastrophe.

One reason it is difficult to recognize, in the history of psychiatry, a continuing clash or tension between these two systems of thought—madness-as-dysfunction and madness-as-strategy—is because it is very easy to confuse that distinction with others. It takes extraordinary vigilance, ongoing mental

exertion, and patience to anchor one's perspective firmly in the soil of teleology and dysteleology, rather than mind and brain. But once we do, we
begin to engage quite differently with texts that once seemed familiar, texts
we thought had nothing left to tell us.

III.

What follows is a schematic summary of the trajectory of this book, a rough-
and-ready road map. But any such synopsis of this nature is fraught with
peril. There are at least two dangers. *First*, a schematic overview, because of
its brevity, might inadvertently suggest, or insinuate to the unwary reader,
the false idea that these two different frameworks chronologically alternate
with each other: in the beginning, there was madness-as-strategy, then there
was madness-as-dysfunction, then there was madness-as-strategy again—
and now madness-as-dysfunction. These are not chronologically alternating
frameworks; they are not "paradigms" in the Kuhnian sense; they have always coexisted, perhaps with one, now the other, having a slight edge in scientific and lay consciousness.

Moreover, it would be an error to suggest that we could even divide up
specific thinkers into two disjoint sets in any absolute manner: those who
endorse madness-as-strategy and those who endorse madness-as-dysfunction. It would be a mistake, for example, to say, *simply*, that Goldstein and
Freud were proponents of madness-as-strategy, but Kant and Haslam were
proponents of madness-as-dysfunction. In most thinkers, we see some inter-
mingling of teleology and dysteleology. Even Kant, our paragon of madness-
as-dysfunction, recognized, in some forms of madness, the wise providence
of nature. Madness-as-strategy and madness-as-dysfunction must be seen as
two tendencies of thought that overlap, not only in one and the same era but
often in one and the same thinker.

Finally, madness-as-strategy is not a *tradition* in the proper sense of the
term. It is not a set of teachings passed down faithfully from generation to
generation. We cannot trace a chain of influence from thinker to thinker,
healer to healer. Rather, what we find are texts and thinkers that resonate with
one another though they emerged at quite different times and places: a spiritual brotherhood. *It is just as if* there is a kind of tendency or need, implanted
deep in the human mind, that must find purpose and goal-directedness even
where it would seem least likely to occur. When Hamlet's Lord Polonius

quipped, "Though this be madness, yet there is method in't," he gave voice to this tendency to see a hidden purpose, a secret end, in what would otherwise strike us as pure psychic pandemonium. But to say that madness-as-strategy is the product of a psychological tendency to see purpose and goal-directedness cannot be read as an attempt to *debunk* it, that is, to dismiss madness-as-strategy as a byproduct of a "psychological mechanism." Madness-as-dysfunction, too, injects an equal measure of purpose and goal-directedness into the world because it sees in madness *so many deviations* from this purposiveness. Both madness-as-dysfunction and madness-as-strategy are indissociably locked into a teleological vision of the world.

There is a *second* danger in such a synopsis of the book's contents: an overview of this sort is bound to be far too condensed, too abbreviated, to serve as a working road map, and, in fact, to serve much good at all. It is Aristides' "speech that says everything all at once." Its only justification, if it can be justified, is to reassure the reader of the existence of a trajectory of thought, that each chapter has a reason for being. The beginning section of each of the book's three parts gives a somewhat more detailed overview of the contents of the section. Still, there is no synopsis or abbreviation or précis that can possibly serve as a *substitute for reading the book.*

Now for the synopsis. In classical times, the Hippocratic physicians excoriated the mages, conjurors, and purifiers, who thought that all forms of madness implemented a strategy—not a human strategy, but a divine one. For the magicians, all forms of madness were counted as so many different *punishments for sin.* As with Nebuchadnezzar, madness is a divine recompense for actions that cause offense to the gods. The key to healing the mad, then, was to *discover which god the patient had offended and to quickly make amends.* The Hippocratic physicians, in sharp contrast, recommended the madness-as-dysfunction model, in which the various forms of madness are simply so many ways that the parts of the body can err: the flow of air could be blocked or the humors could be corrupted or thrown out of proportion to one another.

In the Middle Ages and Renaissance, this teleological and theological portrait of madness was *Christianized*; the physicians, exorcists, and witch-hunters of the day were able to witness, in the mad person, the unmistakable hand of God. But this teleological orientation was now placed inside of a Christian framework, one in which God's justice is always tempered by his mercy. I call this medieval vision the *dual teleology of madness* because madness is never *simply* a punishment for sin. Madness is also, at the same

time, an opportunity for salvation and for the continuing labor of sanctification. It is a tool for attaining a divine end and a manifestation of God's un-
yielding love.

In Robert Burton's *The Anatomy of Melancholy* of 1621, this dual teleological structure is present, but it begins to adopt a naturalistic guise: God has
woven this requirement for punishment into the causal order of the world.
In other words, madness is an inexorable consequence of the misuse of one's
God-given faculties. Various forms of madness are both consequences of
and hyperbolic portraits of their corresponding sins. George Cheyne also accepted this dual teleological structure—madness as both punishment and
means of salvation—but the sin that loomed largest in his thinking was the
sin of rejecting God's generous provision. Melancholy, for Cheyne, is a consequence, first and foremost, of gluttony and intemperance, *but gluttony
and intemperance are, he insists, themselves manifestations of a deeper rebelliousness against God, and specifically, a rejection of God's place-based providence.* The reason for the prevalence of melancholy amongst the English is
their damnable need to import heavy foods, rich wines, and abundant sauces
and spices from the four corners of the earth, imbibements that God never
intended the English to enjoy. In the best of cases, God uses melancholy to
draw our attention to the rebelliousness of our hearts and, ultimately, to lead
us to submit to his sovereignty.

With Kant, that Hippocratic vision, madness-as-dysfunction, reasserts itself forcefully; teleology steps into the background. Kant recognized, in his
youthful essay, only three different basic forms of madness, because there
were only three faculties of the mind and each form of madness is nothing
more than a breakdown or failure of one of those faculties. Madness no longer
had a truth of its own to reveal. By the time of his mature *Anthropology*, Kant
came to recognize six different forms of madness, but that is because, by the
end of the century, he came to recognize *six* different faculties, rather than
three—and hence six different ways the mind could err.

John Haslam, apothecary to Bedlam hospital at the beginning of the nineteenth century, shared this Kantian vision. For him, when madness happens,
it is because one or more faculties (imagination, memory, judgment) fail to
perform their corresponding function: "The sound mind seems to consist
in a harmonized association of its different powers. . . . The different forms
therefore under which we see [insanity], might not, perhaps, be improperly
arranged according to the powers which are chiefly affected."[3] By the nineteenth century, madness-as-dysfunction had come to represent a significant

organizing principle for psychiatry. Nineteenth-century textbooks in psychiatry, like Spurzheim, Cox, and Rush, routinely begin by enumerating the faculties of the *sound mind*, and then demonstrating how the forms of madness represent the breakdown of these faculties. In order to treat a mad person, the only information one absolutely needs is this: which faculty of his mind has erred?

Still, even in the nineteenth century, this teleological impetus makes itself felt; we continue to see the urge or tendency to find purpose and goal-drivenness buried deep inside madness. (The mind *must* see purpose in the world because the mind accords with reality and reality has a purposeful structure.) Philippe Pinel, who was famously said to have broken off the chains of the mad patients at the Bicêtre, recognized in some forms of madness an implicit striving, a kind of mysterious vital principle in nature. He wondered whether some psychotic episodes, which he called *accès de Manie*, were not salutary and therapeutic, whether they did not represent the "salutary efforts of nature [*efforts salutaires de la nature*]."[4] For Arthur Wigan, at least some forms of madness happen when reason submits to madness, as it were, because madness promises something it cannot otherwise have. Johann Christian August Heinroth, too, saw in at least some forms of madness a *coping strategy*, an attempt to flee the pain, rejection, and humiliation of reality. This form of flight involves the creation of a delusional world, and in the best of cases, nature uses this delusional episode to bring about a renewal of the mind. Even Griesinger, who ushered in the era of what historians call "German imperial psychiatry," and who generally thought of the forms of madness as so many dysfunctions of the nervous system, was still able to see, both in dreams and in systems of delusion, the operation of a wish-fulfillment mechanism. Griesinger exemplifies the way in which madness-as-strategy and madness-as-dysfunction are often interwoven within one and the same thinker.

With Freud we arrive at a fundamental, and unabashedly teleological, reorientation of the study of madness, in which its core forms—hysteria, obsessional neurosis, delusions—represent strategies for the deviant fulfillment of unconscious wishes. These strategies serve the dual function of fulfilling, partially, the unconscious wish, while at the same time preventing its content from being consciously recognized. These forms of madness are goal-driven and goal-directed. This teleological reorientation was carried further by Frieda Fromm-Reichmann and Harry Stack Sullivan, who attempted to bring psychoanalytic insights to the treatment of psychotic patients, and it was ultimately incorporated into the first edition of the *DSM*, which describes the three main forms of non-organic mental disorders—the psychotic disorders, the neurotic disorders, and the personality disorders—as *three different strategies for coping with stressors*. For a very brief time, madness-as-strategy was canonized in official psychiatric nosology.

Goldstein, too, helped to carry out this teleological reorientation of psychiatry, but he *biologized* this Freudian approach; instead of conceiving of madness in terms of the relationship between the conscious and unconscious parts of the mind, he understood it as a means by which the organism adapts itself to the environment. Mental disturbance, he thought, *always* expresses the essential organismic tendency toward self-actualization; it is neither an interruption nor an impairment in the ongoing project of self-actualization. With Laing and the anti-psychiatrists, madness assumes a new, social function: the mad person is "deoedipalized"; she is radicalized; she becomes an agent of political resistance, a great "no" to fascism.

Finally, the second edition of the *DSM*, of 1968, renews the Kantian impulse; it begins the long process of eliminating reference to teleology from the *DSM* franchise, and imposing *dysfunction* as its organizing concept instead. First, the *DSM-II* expunged reference to schizophrenic "reactions," and it downplayed the idea that the three main forms of non-organic madness were coping strategies. The *DSM-III* of 1980 brought this eliminative process to completion by abolishing reference to the "neuroses" entirely and declaring, outright, that all mental disorders involve "behavioral, psychological, or biological dysfunction[s]." *We must be very cautious here*: the *DSM-III* had nothing to do with *biologizing* mental disorder, as some philosophers and historians have had it; rather, it had everything to do with putting it into a *dysteleological* framework. The *DSM-III* was not the victory of the biological orientation but the victory of madness-as-dysfunction. Today's RDoC is the inevitable fruition of that movement.

At present, we find ourselves at a curious impasse. On the one hand, madness-as-dysfunction appears to be stronger than ever; we have attempted to identify, and in some cases perhaps even succeeded in identifying, dysfunctional neural circuits and ultimately dysfunctional gene mutations underlying some forms of mental illness. On the other hand, we are also entering the era of *Darwinian medicine*, which is just as likely to see specific forms of disorder as Pleistocene adaptations, that is, as "Darwinian algorithms," as it is to see them as evolutionary dysfunctions. In the era of Darwinian medicine, we can now question whether depression is an evolved strategy for navigating interpersonal conflict, or for detaching from unrealistic life goals; we can wonder whether some of the anxiety disorders are "life-history strate-gies" for remaining vigilant to potential threats in hostile environments; we can even ask whether the delusions of schizophrenia are not strategies for coping with perceptual abnormalities. Hence the end joins the beginning: we can see a set of resonances that fuse together the ancient Greek magicians, the exorcists and witch-hunters of the Renaissance, the psychoanalysts, and the evolutionary psychologists: all of them willing to see, in madness, the working out of a strategy, purpose, or end.

IV.

I have attempted to draw out, from the theorists of madness, a different way of seeing madness and the forms that it takes, different from the "medical-ization" that is alternately praised and reviled, and which I will define here, in a highly provisional manner, as the way of seeing that takes as its starting point for approaching madness, *function and its deviations*, and all that it entails, in terms of observation, manipulation, intervention. This alternative way of seeing proceeds from the conviction that some people are truly *mad by design*, that at least some of its forms are strategies for solving problems, coping with aspects of the environment, regulating one's mental economy. More importantly, I have sought to demonstrate that theorists and thinkers throughout the millennia have adopted this way of thinking as a starting point, and that for *them*, the idea that madness is *not* a strategy, that we *cannot* see in it the creative problem-solving activity of the human mind, or God's stern but benevolent hand, or the healing impetus of nature, or an ev-olutionary adaptation, a purpose-driven process, a well-operating machine, borders on incomprehensibility.

But what is the value of this recovery operation? What, indeed, is the virtue of reorganizing our thinking about madness in terms of teleology and dysteleology, rather than mind and body? There are four such virtues. First, while this is not a work of history of science, the distinction between madness-as-dysfunction and madness-as-strategy provides a new set of tools to the historian; it gives the historian new ways to tell stories about the past. It yields novel configurations of actors (the ancient Greek magicians and purifiers, Pinel, Freud, and the evolutionary psychologists form a single block). It illuminates specific controversies or debates (the clash be-tween Pinel and Haslam on the usefulness of autopsy was not a question of mental versus physical, but one of strategy versus dysfunction). It helps us to reconstruct major transitions in that history (for example, the shift from *DSM-II* to *DSM-III* was not a shift from a psychogenic to a biogenic manual, as some historians have offered, but a shift from madness-as-strategy to madness-as-dysfunction).

Second, and more important, is its connection to the *philosophy* of medicine and psychiatry. One orthodox view about the nature of mental illness is the dysfunction-centered position of Wakefield and Boorse, who think that, as a matter of *conceptual analysis*, diseases (including madness) involve dysfunctions or failures of function. In other words, in the late twentieth century, philosophers converted the *historical* entrenchment of madness-as-dysfunction into a matter of *logical necessity*. Recasting the history in terms of this dialogue can break open the fixation with dysfunction among the philosophers. Crucially, *I do not offer here a new definition of madness*. It is not a question of setting definitions against one another: *this one* prefers the harmful dysfunction definition; *that one* prefers the biostatistical definition; *yet another* prefers the madness-as-strategy definition; *still others* prefer a family resemblance definition, and now we shall evaluate the merits of these definitions, examine their goodness of fit with existing biomedical and psychiatric usage, construct ingenious thought experiments to weigh in favor of one or the other. I present not a definition but a mode of consciousness, a *Denkstil*, that, intellectually speaking, both predates any particular definition and conditions the apparent plausibility of that definition. One who accepts madness-as-dysfunction, who is anchored in this special way of seeing, will view certain philosophical definitions, such as those of Boorse or Wakefield, as containing an inherent plausibility, as overwhelmingly persuasive, perhaps even as inevitable. In fact, as Fleck showed, it is possible for a style of thought to have such a firm grip on the mind, that any departure from it will appear inconceivable.[5] But this inconceivability must not be taken as a sign of its status as a necessary truth, but as a diagnostic tool for measuring the degree of one's entrenchment in this way of seeing.

A third potential benefit has to do with research and treatment itself, that is, the manner in which people are healed, in those cases where healing is either desirable or necessary. If I think that your illness, or even an aspect of it, is an adaptive, purposeful or goal-directed response to a problem, I am less likely to attack it as if it is the problem itself. Madness-as-strategy helps to modulate the logic of intervention. As classicist Ludwig Edelstein noted, "In later centuries [that is, following antiquity] the recognition of nature as a teleological power must have confirmed the advisability of the withdrawal of the physicians."[6] Edelstein's phrasing is far too strong, but there is a ring of truth in his comment. If the delusions of schizophrenia have an adaptive or functional role in compensating for perceptual abnormalities, or for yielding an appearance of meaning in a seemingly absurd or cruel world, then one would not want to target the delusions as if they represent the pathology itself.[7] One must rather target the situation or event or arrangement, whether inner or outer, to which the delusions are an adaptive response.

Finally, this recovery operation, this attempt to retrieve madness-as-strategy as a coherent way of seeing, contributes to the project of providing *intellectual scaffolding* for the emerging movement variously known as Mad Pride, mad resistance, or mad activism.[8] This movement is helmed by mental health service users (ex-patients, survivors) who demand, above all, that society adopt new forms of conceiving of, and engaging with, madness, who insist that madness, far from being a deficit or pathology, has a truth of its own to reveal, that it has a *positive being* and not simply the being of a negation, lack, or failure. The very terminology of "mental illness" or "mental disorder" is, as a rule, shunned because such language attempts to strip madness of this positive being; it places madness on a shelf as a kind of medical curio, along with a two-headed fetus or oversized tumor. The mad are not a curiosity; madness is not always a disease to be cured but a force of disruption to be reckoned with. We must view madness in its destabilizing and transformative tendency; in the final analysis we do not "view" madness at all, as if we stand in a subject-object relation to it, as if we contemplate this thing, this *phenomenon*, from a safe distance; we must be taken up into the air by madness and deposited onto a new soil, a land of beautiful and terrifying foliage, of wonderful animals and spirits, a land both strange and familiar, as familiar and as perplexing as one's very hand. In the end, madness is not a specific mode of existence of the person but the default state of humankind, ground zero of the conscious mind.

PART I

THE DUAL TELEOLOGY
OF MADNESS

Before the eighteenth century, madness could scarcely be conceived outside of the context of teleology and purpose. That is because madness was nearly always understood, as was disease generally, as a pivotal component of God's purpose for humanity as a whole. The basic question that practically all theorists of madness wrestle with is, *how does madness stand with God*? And this should not be understood merely in terms of a boring theodicy: *why would a good God allow such torments to befall his creatures*? But rather: *how does madness, and disease more generally, serve God's ultimate ends*? Madness is always by design; madness is always a strategy, a divine one. God uses madness to reform our hearts and cleanse our minds.

The real danger for *us*, *today*, is that in burying this theistic vision of madness, in blithely dismissing this theological orientation toward the mad, we also, at the same time, *bury teleology along with it*. That would be an egregious error, an unforgivable sin. Teleology can survive the "death of God" that we moderns have heard so much about: teleology is *protean*, it is robust and vigorous because of its ability to assume new forms in new intellectual environments. It can take the form of a divine mandate; it can take the form of a mysterious vital principle in nature; it can take the form of an unconscious idea driving toward fulfillment; it can take the form of the goal-directedness of the organism; it can take the form of a Darwinian adaptation. Some biologists might balk at the marriage of Darwinism and teleology, but we must remember that the *teleological* can be pulled apart from the conscious or even unconscious idea, as will soon be shown.

This picture of madness as divine strategy stretches at least as far back as Hippocrates' time. The Hippocratic physicians are celebrated today for their supposedly naturalistic approach to medicine; we are often reminded that the author of *On the Sacred Disease* rebukes the magicians, the conjurors, the purifiers, who think of madness as a form of divine punishment, and its treatment as demanding a swift propitiation to the angry god. Our story, then, must begin with those magicians, conjurors, purifiers. Ultimately, the Hippocratic doctors were not opposing a natural vision of medicine against their supernatural vision. Rather, they were opposing a vision of medicine whose fundamental explanatory principle was the principle of *dysfunction* against one whose fundamental principle was that of *strategy*. In a sense, this entire book can be read as *a long-overdue defense* of the conjurors, the magicians, the purifiers.

By the Middle Ages and the Renaissance, we are fully nestled in a Christianized vision of the world. Madness is still a strategy, but now it is, even more clearly, a strategy in a two-fold sense: *it is not just a punishment for sin, but also the means of salvation and the ongoing work of sanctification.* We have left the harshness of the classical world; we have entered a world the overriding theme of which is God's love, God's *longing*, for the salvation of the individual and her continuing sanctification, mediated by his word and his indwelling spirit. Madness always exemplifies, fuses, these two characteristics of God: God's justice and God's mercy. This is what I call the dual teleology of madness. This dual teleology, however, is construed differently by different thinkers.

In the fifteenth and sixteenth centuries, in the debate between the physicians and the exorcists, we see a mirror of the earlier debate between Hippocrates and the magicians. For the exorcists, many of the cases that the physicians deem madness are, in fact, cases of demon possession. But, they insisted, God has allowed this person to become possessed for the twin purposes of punishment and redemption. In opposition to the exorcists, witch-skeptics like Jorden, while acknowledging the dual teleology of madness, nonetheless insist that madness must be understood, first and foremost, in terms of dysteleology: in his case, in terms of the organic dysfunction he calls *suffocation of the mother*, or hysteria. We witness Jorden, then, attempting to institute madness-as-dysfunction *inside* madness-as-strategy, attempting to carve out a little space for dysteleology inside a world infused with purpose, a world in which all things march toward their natural end.

For Robert Burton, in the early seventeenth century, the dual teleology of madness is axiomatic—all disease has the role of punishing and purifying— but he begins to *naturalize* this divine strategy. For Burton, madness does not *always* require a special intervention on God's part. When someone is mad, it is not because God has thought, "I see that you have sinned, and I am sending madness to you, to punish you and hopefully to redeem you." Rather, God has designed the material universe in such a way that certain kinds of sins will *inexorably* be followed by certain kinds of punishments and, along with punishment, the opportunity for redemption: think gluttony and diabetes, or heavy drinking and liver damage. When God made our bodies, he "did not need to do anything else," as the philosophers like to say, to ensure that certain sins would be visited by certain forms of madness.

Finally, George Cheyne, in his 1833 classic, *The English Malady*, also understood madness, first and foremost, as a revelation of God's purpose for humanity. Diverse diseases such as hysteria, hypochondria, and melancholy, for Cheyne, come about through nerve damage. Nerve damage, however, is born of intemperance, and intemperance, in turn, of our rejection of God's place-based providence. Madness, therefore, is as much antidote as disease: it is *God's antidote to the madness of society.* The misery of melancholy is God's way of inviting us to conform our lifestyles to his original plan, given the region of the globe on which he has placed us. With Cheyne, we also begin to see a certain inversion of thought that becomes a cornerstone of the anti-psychiatry of the 1960s: what society calls "sanity" is, in truth, a form of collective madness; the "mad person," in resisting *so-called* sanity, is the one that is, or is on the road to becoming, truly sane.

1

Hippocrates and the Magicians

We begin with the Hippocratic physicians and their clash with the "magicians, purifiers, charlatans and quacks." What, precisely, is the nature of this clash between Hippocrates and his opponents? A rather obvious and—to contemporary ears—almost irresistible way to understand the clash is in terms of *naturalism* and *supernaturalism*, as advocated most recently by Jacques Jouanna (Section I). Unfortunately, this reading, at least in a simplistic form, has not survived the rigors of classical scholarship: as Ludwig Edelstein and G. E. R. Lloyd emphasize, the Hippocratics, too, invoke divinity, purification, and prayer, just as their opponents do (Section II). Here, I suggest—with all of the intellectual humility the topic demands—an alternative way of reading the opposition between Hippocrates and the magicians: the relevant distinction is that between *dysteleology and teleology* (Section III). The Hippocratics were, by and large, wedded to the idea that *when madness happens, something has gone wrong inside of you; something within you fails to serve its purpose or function.* Disease is the breakdown of teleology. For the magicians and purifiers, in contrast, all disease, including madness, *carries teleology inside of it*; disease has a divine purpose and justification. It is in this sense that this book can be read as a long-overdue apologia for the magicians and purifiers.

I.

Medicine begins on the island of Kos, around 400 BC, with the physician Hippocrates and his followers. This, at least, represents the rough consensus of historians of medicine today.[1] This is not to say, however, that the *healing arts* themselves begin around 400 BC in Kos; healing is universal. Possibly as early as 600 BC, Asclepius was regarded as the patron saint of healers, and healing was practiced in some form or another throughout the Near East.[2] So, in what sense does medicine "begin" with the Hippocratic physicians? Why does Hippocrates deserve the honor of being the father of medicine? In what way did the Hippocratic physicians deviate from other practitioners?

Madness. Justin Garson, Oxford University Press. © Oxford University Press 2022.
DOI: 10.1093/oso/9780197613832.003.0002

What is the *principle*, if there is, in fact, a principle, that radically marks off the Hippocratic style of doing medicine?

One way to answer this question is to interrogate the Hippocratic text itself: what does Hippocrates—taking "Hippocrates" to refer to a large collection of texts, the bulk of which were probably penned in the fifth and fourth centuries by various writers, rather than an individual man—what does Hippocrates itself, thus understood, say about its *own* originality, its *own* deviation from the norm? What is Hippocrates *against*? What does Hippocrates set itself up in opposition to?

We can use *On the Sacred Disease* as a fundamental source here. It is particularly apt because of its prominent polemical function; one of its jobs is to contrast, repeatedly, this new style of doing medicine with a motley assortment of healing traditions.[3] The author of *On the Sacred Disease* tells us that this new way of practicing medicine stands in sharp contrast to the way of the "magicians, purifiers, charlatans and quacks."[4] So, right at the outset, we know what the Hippocratic practitioner is *not* about: he is not about magic, incantations, and other superstitious nonsense.

Who are these sordid figures, these charlatans and quacks? What more can be said of them? What, precisely, are the practices they engage in that the Hippocratic author finds so detestable? While we cannot know with certainty, we can assemble some of the Hippocratic texts to construct a composite portrait of this earlier tradition that the Hippocratics condemn. Consider, once again, *On the Sacred Disease*:

> Those who first attributed a sacred character to this malady were like the magicians, purifiers, charlatans and quacks of our own day, men who claim great piety and superior knowledge. Being at a loss, and having no treatment which would help, they concealed and sheltered themselves behind superstition, and called this illness sacred. . . . They used purifications and incantations; they forbade the use of baths, and of many foods that are unsuitable for sick folk.[5]

And:

> By these sayings and devices they claim superior knowledge, and deceive men by prescribing for them purifications and cleanings, most of their talk turning on the intervention of gods and spirits.[6]

Here, then, is an obvious way, an almost *irresistible* way, of grasping the originality of the Hippocratic corpus: Hippocrates teaches us to practice medicine
using *natural*, rather than *supernatural*, means. The Hippocratic physician
practices medicine using natural means and conceptions alone, both in etiology and in treatment. His opponent practices medicine by drawing upon
supernatural means and conceptions, both in his explanations and in his
cures. And no doubt there is some truth in this, that is, in invoking the distinction between natural and supernatural to capture the distinguishing
mark of the Hippocratic text. Consider, for example, the way our opposing
groups of practitioners explain epilepsy, the so-called sacred disease. A magician wants to cure epilepsy. He assumes, at the outset, that the patient has
done something to offend the gods. The urgent question that must be
answered, then, is this: which god has he offended? This is what "differential
diagnosis" consists of:

> If the patient imitate a goat, if he roar, or suffer convulsions in the right side,
> they say that the Mother of the Gods is to blame. If he utter a piercing and
> loud cry, they liken him to a horse and blame Poseidon. . . . If he foam at the
> mouth and kick, Ares has the blame. When at night occur fears and terrors,
> delirium, jumpings from the bed and rushings out of doors, they say that
> Hecate is attacking or heroes are assaulting.[7]

In contrast, consider the corresponding Hippocratic explanation of
symptoms of epilepsy. Epilepsy is caused by air that is trapped in the veins
due to the accumulation of phlegm, and which cannot move about freely. The
different symptoms of the sacred disease occur when different parts of the
body are deprived of air:

> The hands are paralyzed and twisted when the blood is still, and is not dis-
> tributed as usual. The eyes roll when the minor veins are shut off from the
> air and pulsate. The foaming at the mouth comes from the lungs; for when
> the breath fails to enter them they foam and boil as though death were
> near. . . . The patient kicks when the air is shut off in the limbs, and cannot
> pass through to the outside because of the phlegm; rushing upwards and
> downwards through the blood it causes convulsions and pain; hence the
> kicking.[8]

Later, the same author describes *all* forms of madness as the result of excessive heat, cold, dryness, or moistness, in the brain:

> Those who are mad through phlegm are quiet, and neither shout nor make a disturbance; those maddened through bile are noisy, evil-doers and restless, always doing something inopportune. These are the causes of continued madness. But if terrors and fears attack, they are due to a change in the brain. Now it changes when it is heated, and it is heated by bile which rushes to the brain from the rest of the body by way of the blood-veins. . . . The patient suffers from causeless distress and anguish when the brain is chilled and contracted contrary to custom.[9]

The opposition between natural and supernatural not only appears to capture the originality of the Hippocratic tradition, but also provides a philosophical *justification* for that tradition. *That* Hippocrates practices medicine naturalistically would seem to justify Hippocrates' pride of place as the father of medicine, at least to modern ears. We live, today, in what Charles Taylor calls a *secular age*; we live in the era of naturalism; we expect our physicians to rely exclusively on natural causes and cures. Even though the Hippocratics were, *empirically speaking*, mistaken in their understanding of epilepsy—we know, of course, that epilepsy stems from uncontrollable electrical discharge in the brain, not trapped air—we, the medical practitioners of today, can still count ourselves as their descendants; we are still their offspring; we still practice medicine within their lineage. I am the seed of Hippocrates if I practice medicine under the aegis, the seal, of Hippocrates' naturalistic approach to health and disease.

II.

This elegant story, while beautiful in its simplicity and in its resonance with modern, secular sensibilities, cannot be sustained. The distinction between the natural and the supernatural cannot, in any absolute way, capture whatever is supposed to represent the decisive break of the Hippocratic tradition. This is because the Hippocratic doctor, too, appeals to the supernatural, just as the magician appeals to nature. Classicists Edelstein and Lloyd have done much to help to reorient our thinking of this history.[10] One early hint of the failure of the distinction between natural and supernatural to capture the specificity of "Hippocrates" is given to us in the opening lines of *On the Sacred Disease*. There, the author says of the sacred disease that "it is not, in my opinion, *any more divine or more sacred than* other diseases."[11] He does not say that it is *not* sacred; rather, it is not *more* sacred than others. Later, he clarifies his position: "Quotidian fevers, tertians and quartans seem to me to be no less sacred and god-sent than this disease, but nobody wonders at them."[12] Finally, as if to remove any doubt about the divine origin of disease, the writer affirms that *all diseases are equally divine*:

> This disease styled sacred comes from the same causes as others, from the things that come to and go from the body, from cold, sun, and from the changing restlessness of winds. These things are divine. So that there is no need to put the disease in a special class and to consider it more divine than the others; they are all divine and all human.[13]

One might think that such passages represent a slip or a loose manner of speaking, but they do not. Later, the author accuses the magicians of impiety because of the detestable *manner* in which they invoke the supernatural. To the extent that your suffering is caused by offense against the gods, you ought to bring yourself at once to the temple and make the proper purifications there. There is no question that the gods are intimately involved in health and disease:

> In making use, too, of purifications and incantations they do what I think is a very unholy and irreligious thing. For the sufferers from the disease they purify with blood and such like, as though they were polluted, bloodguilty, bewitched by men, or had committed some unholy act. All such they ought to have treated in the opposite way; they should have brought them to the sanctuaries, with sacrifices and prayers, in supplication of the gods. As it is, however, they do nothing of the kind, but merely purify them.[14]

Such a passage would have no place in a text devoted to extolling a new, naturalistic vision of medicine. The author does not object to the proposition that disease has a divine origin; he objects to the crude and superstitious way in which the magicians carry out their sacred rites. The relevant contrast here is between religion and magic, not between natural and supernatural.[15]

This peculiarity, this Hippocratic invocation of the divine, is not confined to *On the Sacred Disease*. Consider another Hippocratic text: *Airs, Waters,*

Places, probably written by the same author. There, we are told, "But the truth is, as I said above, these affections [impotence] are neither more nor less divine than any others, and all and each are natural."[16]

Not only did the Hippocratic authors nod, in principle, to the possibility that disease has a divine source, but another Hippocratic text, *Dreams (Regimen IV)*, advocates that prayer be directed to different gods depending on the state of the body. Here, the author shows us how dream images hold the key to understanding the origins of disease:

> So with this knowledge about the heavenly bodies, precautions must be taken, with change of regimen and prayers to the gods; in the case of good signs, to the Sun, to Heavenly Zeus, to Zeus, Protector of Home, to Athena, Protectress of Home, to Hermes and to Apollo; in the case of adverse signs, to the Averters of evil, to Earth and to the Heroes, that all dangers may be averted.[17]

It is extremely tempting at this juncture to read these references to the divine in the Hippocratic text as mere *vestiges* of an older tradition that had yet to be entirely expunged, or as a result of the *polyvocality* of the Hippocratic tradition. Surely, one might argue, the Hippocratic tradition did not spring into being in a day. Some of Hippocrates' followers must have been converts from these earlier, magico-religious traditions, and some of that superstitious thinking remained attached, as an ugly residue, to the Hippocratic corpus, despite the fact that it is, in its very essence, alien to that corpus. Edelstein, for example, notes how scholars, he thinks mistakenly, dismiss Hippocratic references to magic and religion as "rather of the drapery than of the body of medicine."[18] And Jouanna downplays the significance of this passage from *Regimen IV* by saying that it "occupies a relatively exceptional place in the Hippocratic corpus for its position regarding the sacred," and that *On the Sacred Disease* and *Airs, Waters, Places* are the "most representative" treatises.[19]

Jouanna also emphasizes, in regard to these problematic passages, the extent to which the Hippocratic corpus is itself *polyvocal*: "Despite a common element of rationalism that unites the treatises, it would be futile to suggest that all the Hippocratic doctors held a unified position on the sacred."[20] The suggestion here is that the passage in *Regimen IV* is a trace of another tradition that was on its way to extinction; it is not proper to the Hippocratic tradition; it has insinuated itself there without permission; it is our job, as historians and philosophers and classicists, to discern, within the Hippocratic corpus, what is representative of that tradition and what is foreign to it, and to expel the foreign elements.

And perhaps Jouanna is right; maybe we should see, in these passages, remnants of an earlier tradition that had not yet been thoroughly stamped out; perhaps these passages were written by people who had not entirely entered into the inmost spirit of the Hippocratic philosophy. But what of the references to the divine (θεια) in *On the Sacred Disease* and *Airs, Waters, Places*? Jouanna hastens to assure us that the term, here, refers not to an anthropomorphic deity, but to the law-governed realm of nature as a whole, to an abstract and thoroughly impersonal principle of order. Commenting on the opening lines of *On the Sacred Disease*, he writes, "Instead of a more or less obscure divine justice, which punishes the guilty with a disease, the Hippocratic doctor proclaims a universal order that is both divine and natural, which accounts for all disease and frees the patient from all guilt"[21]— an interpretation that surely stands in tension with the later passage recommending that the sick be brought "to the sanctuaries, with sacrifices and prayers, in supplication of the gods." It also stands in tension with Jouanna's observation that "the rationalism of the Hippocratic doctors was not incompatible with acknowledgement of the traditional gods and participation in the cult of the great sanctuaries."[22] And it stands in tension with the opinion of earlier scholars, such as Edelstein, who insist that θεια refers to a supernatural realm and who see, in the opening of *On the Sacred Disease*, an admission that "the two spheres of the divine and of the natural are then fundamentally separate, although their influence is combined in every action."[23]

Jouanna's hermeneutic strategy is necessary, but also dangerous. All philosophers, historians, classicists, must engage in "rational reconstruction." If I am reading a text of Hegel, I must be willing to declare certain passages to be unrepresentative of the *true*, or *mature*, or *authentic* Hegel. This is *necessary*, for no text is perfectly self-consistent. But, in applying this strategy—and here is the danger—there is always the risk that, in our reconstruction of a certain text or thinker, we deny the complexity of their thinking, and we project back onto that thinker a mirror of our contemporary creeds. Consider how scholars have attempted to "reconstruct" Darwin, how they have dismissed or marginalized passages that do not neatly fit into our current understanding of evolutionary biology, such as the extent of his reliance on the inheritance of acquired characteristics, or the role of group selection in his treatment of biological altruism. The danger in Jouanna's dismissive attitude, then, is that it might overlook or ignore the extent to which the divine is woven into the Hippocratic corpus, not as an abstract philosophical principle of Order, but as a *being*, both like and unlike myself, with thoughts, feelings, preferences, and urges, and who will not hesitate to destroy my health and livelihood in his fury.

One Hippocratic text that contravenes this dismissive attitude is the *Oath* itself, certainly, by today's reckoning, the most popular of the Hippocratic texts, even if not as highly regarded by the ancient Greeks. The *Oath* begins with an invocation to the gods:

> I swear by Apollo Physician, by Asclepius, by Health, by Panacea and by all the gods and goddesses, making them my witnesses, that I will carry out, according to my ability and judgment, this oath and this indenture.[24]

Before we rush to dismiss the invocation as flowery rhetoric or as an empty formalism, we must recognize that the *Oath* has the structure of the ancient Near East covenant, examples of which are found in Mesopotamia, Anatolia, and the Levant, a poignant example being Deuteronomy 28: if I do X, may the gods bless me; if I fail to do X, may they curse me.[25] One does not take such a covenant lightly. The function of the covenant is to set out the terms and conditions of my conduct with regard to a certain venture. If I satisfy those terms and conditions, I invite the gods to bless my life and career; if I violate those terms and conditions, I draw down curses upon my own head:

> Now if I carry out this oath, and break it not, may I gain for ever reputation along all men for my life and for my art; but if I transgress it and forswear myself, may the opposite befall me.[26]

Now, the very act of taking such an oath raises a deep puzzle. Suppose I am a physician in the "Hippocratic tradition" who has taken this oath, and earnestly so, not as an empty formalism or pretention, but as if my very livelihood depended on it. Suppose I later violate the terms of this covenant: I have sex with my patient's servant while my patient is comatose; I divulge gossip about my patient to my colleagues; I neglect the right medicine; I offer an abortion to a terrified teen. Thus I invoke the wrath of the gods on my "life and art." *To whom do I turn for healing?* Certainly *not* my Hippocratic colleagues, who know how to dissipate the overabundant or corrupt phlegm and choler, but who have forgotten how to pacify the gods that my unjust actions have angered. Whom else can I turn to, but the temple purifiers? Hence these two traditions, this "naturalistic" Hippocratic tradition, and this "supernatural" tradition, this tradition that sees the role of deity in health and disease, cannot be seen as fundamentally opposed, not only because of this interpenetration of methods and ideas, *but because the divine stands behind the Hippocratic practice as a guarantor*, as underwriter.

III.

I want to recommend a different distinction for grasping what is specific to this Hippocratic rupture. This is not to deny, in toto, the role of the natural and supernatural in thinking through this rupture. But another distinction reveals a new facet to this break: teleology and dysteleology. Return to the Hippocratic explanation of epilepsy. We have the organism, the body with its various organs, and each organ has its own little job to do, its role to perform, its special purpose. Disease happens when these organs, the organism's parts and processes, fail to do their jobs. They fail to carry out their purposes. They cannot do what they are supposed to do. If teleology can be likened to an arrow hurtling toward a target, disease happens when the arrow misses its mark.

In short, what is specific to the Hippocratic rupture is this: disease originates from a disruption or violation of teleology. It is the negation of teleology. Disease has no goal, no purpose, no end, and it perverts the creature's ability, and the ability of its parts, to carry out *their* jobs. Disease frustrates teleology at every level of biological organization. And, to the extent that the *purposeful* is the *good*—to the extent to which a creature's good is found in its parts achieving their ends—then medicine is, necessarily, good, because it restores the creature's capacity to attain *its* good. On the contrary, our magicians, our conjurors and purifiers, think of disease as serving, or as satisfying, a purpose, in this case, the purpose of divine retribution, of just penalty for offense. Disease has a reason for being, a telos, an end, an *in-order-to*. Disease does not represent the frustration of purpose—or not *merely* its frustration—but its satisfaction. What this means for health is that we must discover, through careful observation, which god has been offended, so we know whom to appease. The split between teleology and dysteleology captures something essential in the schism between the Hippocratics and the magicians. Ultimately, the distinction between teleology and dysteleology can be decisively *severed* from the distinction between the natural and the supernatural. It is not as if the teleological aligns itself with the supernatural, and the dysteleological with the natural. In a sense, the purpose of this book is to reveal, precisely, how they come apart, and why their independence, their *apartness*, is necessary for reorganizing the study of madness.

I am not ultimately interested, here, in venturing a thesis about medicine in the ancient world. I wish to identify something like two contrasting traditions that are woven throughout the history of madness. I call the first tradition, this Hippocratic tradition, *madness-as-dysfunction*. I call the second tradition, the tradition of the conjurors and magicians, *madness-as-strategy*, or, equivalently, *madness-as-design*. Madness-as-strategy is *protean*; it resurfaces in many different guises throughout the centuries. My purpose is to follow its movement through time, to point to it, to speak its proper name, and, in some measure, to conjure it.

But this attempt raises a fresh problem. The Hippocratic tradition is a good, proper tradition. It allows one to trace in it, historically, something like a lineage of appropriation. It offers itself up, readily, to the task of the historian of science. Celsus appropriated Hippocrates. Galen appropriated Celsus. Avicenna appropriated Galen. In fact, the possibility of tracing such a lineage *is itself part of the "Hippocratic tradition."* Hippocrates *itself* insists that it *be* a tradition, that its precepts be passed down from mentor to mentee, from father to son. The *Oath* consecrates this mandate that the teachings be passed down faithfully and in the proper manner: the physician swears, "to impart precept, oral instruction, and all other instruction to my own sons, the sons of my teacher, and to indentured pupils who have taken the physician's oath, but to nobody else."[27]

But this alternative "tradition," the tradition of which the charlatans and magicians are representatives, madness-as-strategy, *has no proper lineage*. The priest-physicians had no successor. We find no mandate that a set of precepts or practices be passed on to anybody else. To be sure, we find resemblances, resonances, through history. The exorcists and the witch-hunters of the Christian era were, in a sense, "descendants" of the magicians and purifiers, but not because of a lineage, not because a body of principles and practices was passed down from one group to the other in an unbroken chain. Freud himself sometimes writes of an intellectual affiliation between psychoanalysis and the healing practices of the Asclepian temple.[28] If the madness-as-dysfunction tradition organizes itself into a proper narrative, this *other* tradition, this madness-as-strategy tradition, is like a child that interrupts that narrative from time to time, but each time in a different costume.

There is much more to be said about what this distinction between madness-as-strategy and madness-as-dysfunction amounts to. These two orientations must, at the moment, be treated as labels, as placeholders that will be progressively filled in as we engage with thinkers throughout the centuries. But we can say confidently, and in advance, that whatever this madness-as-strategy "tradition," or *anti-tradition*, amounts to, it is by no means a celebration of the mystical, the irrational, the superstitious. The magicians and purifiers are a part of this madness-as-strategy orientation *not* because they invoke the supernatural, but because they invoke the purposive. Teleology slices through the natural/supernatural divide: we can speak of the purposes of God, of the purposes of an individual human being, of conscious purposes, of unconscious purposes, of the purpose of an organism, of the purpose of an organ. Moreover, in each domain that purpose or teleology attaches to, dysteleology presents itself as well. The possibility of failure, of missing the mark, of going off course, rests quietly inside purpose; the prospect of failure is the necessary correlate of purpose. But all this remains to be shown in a rigorous way.

2

The Suffocation of the Mother

By the Renaissance, madness acquired a new purpose or end, an end that was vaguely prefigured in the practices of the Greek magicians. God allows madness not merely to punish, but to redeem and sanctify. This is the magical transformation that Christianity has wrought. By the sixteenth and seventeenth centuries, this basic understanding of madness is so pervasive that, even in the skirmishes and disputes between the physicians and the exorcists, *nobody contests this dual teleological character of madness, that is, its role as instrument of both punishment and redemption*; it is a common ground; it is axiomatic (Section I). Nonetheless, even within this theistic and teleological point of orientation, physicians still sought to carve out a region in the space of disease over which they could claim exclusive jurisdiction (Section II). Edward Jorden deserves our special attention here because of the way he delicately attempts to carve out, *within the horizon of madness-as-strategy*, a space for madness-as-dysfunction. Without denying their divine origin, he attempts to demonstrate that most forms of madness stem from an organic dysfunction, hysteria or the "suffocation of the mother" (Section III). In this way, the study of madness is gradually unmoored from its theological ground. Yet, strangely enough, even for those diseases that proceed from organic dysfunction, *he never quite banishes, entirely,* the role of the priest, and the need for spells, incantations, amulets, and songs. These still play a role in healing, but only in the manner of play-acting, as a form of theater (Section IV).

I.

When we reach the Middle Ages, madness-as-strategy is *Christianized*. Gone are the soothsayers, the mages, the Aesclepian purifiers and interpreters of dreams. In their place, we now have the exorcists, the witch-hunters, the papists who seek to ward off disease by compiling hierarchies of

Madness. Justin Garson, Oxford University Press. © Oxford University Press 2022.
DOI: 10.1093/oso/9780197613832.003.0003

devils—Belial, Astaroth, Balam—and who attempt to discern which devil is implicated in the patient's complaint, whether it be stone, miscarriage, or frenzy.

One of their sworn opponents was Edward Jorden, physician and witch-skeptic of the beginning of the seventeenth century, famous for his attempt, albeit unsuccessful, to defend Elizabeth Jackson from the accusation of witchcraft. His grounds, in that case, were that Jackson's alleged victim, young Mary Glover, was suffering fits not because of demonic possession, but because of "suffocation of the mother," alternatively known as hysteria, which was natural, and not supernatural. This is how Jorden, in good Hippocratic fashion, reviles his opponents:

> I thought good to make knowne the doctrine of this disease, so farre forth, as may be in a vulgar tongue conveniently disclosed, to the end that the unlearned and rash conceits of divers, might be thereby brought to better understanding and moderation; who are apt to make every thing a supernaturall work which they do not understand, proportioning the bounds of nature unto their own capacities: which might prove an occasion of abusing the name of God, and make us to use holy prayer as ungroundedly as the Papists do their prophane tricks; who are readie to drawe forth their wooden dagger, if they do but see a maid suffering one of these fits of the Mother, conjuring and exorcising them as if they were possessed with evil Spirits.[1]

It should come as little surprise, then, that Jorden, at the outset of the book, explicitly positions himself as the seed of Hippocrates and as laboring under the aegis of Hippocrates' naturalistic method; ultimately, his lonely battle against the exorcists is a self-conscious repetition of Hippocrates' ancient battle against the magicians:

> Wherefore the rule of *Hyppocrates* must needes be true; that if these *Symptoms* do yeeld unto naturall remedies, they must also bee naturall themselves. And thus much in explanation of these two arguments of *Hyppocrates* against the errour of his time: which notwithstanding hath been continued in the minds of men untill this day, and no marvell: *unless the same corruption which bred it at the first, had beene removed out of the world.*[2]

Jorden clearly recognizes that while *he* belongs to a proper tradition, the one established by Hippocrates, the one that assembles itself into a historically continuous lineage, *the exorcists and witch-hunters do not belong to a proper tradition*. Instead, the same superstitious force, the same human tendency, the same *corruption which bred it at the first,* the inner rot that fueled the magicians and purifiers in the time of Hippocrates, *also* fuels the exorcists and witch-hunters of the present. It is as if some innate force of thought, a deep tendency of the human mind, continues to generate superstitious beliefs and practices without anyone needing to teach them to anyone else, without their needing to be passed down from teacher to pupil.

Jorden is wrong in at least one respect: the new situation he militates against, the superstitious rites of the witch-hunters and exorcists, is not *just* a repetition; it is not just a Renaissance incarnation of the practices of the ancient Greek mages and purifiers. The difference is that now, disease has a new purpose, a new telos, one that was only vaguely anticipated in the classical age. The purpose of disease is no longer *merely* punishment for sin; in addition, disease is *a means of redemption*. This structure is so vital to medieval thought about madness that I will call it the *dual teleology of madness*: madness is instrument of punishment as well as salvation and continuing sanctification.

But how does disease play this role? How does it move the narrative of one's life along? Robert Burton, in his *The Anatomy of Melancholy* of 1621, articulates one specialized version of this formula of madness, which he attributes to Pliny: "In sickness the mind reflects upon itself, with judgment surveys itself, and *abhors its former course*."[3] Madness forces a hiatus, a reflection. *How are things with my life?* And that reflection can prompt the turn, or return, to God. This is not to say that *all* disease has the structure of a divine interruption into the path of a wayward soul. For that would imply that God's elect never get sick. If I am already saved, do I need disease to lead me to salvation again? Must I be saved twice? Rather, the more general formula, of which Burton's formula is a specification, is that sickness is, in addition to punishment, a character-forming encounter with the divine, whether it take the shape of an *interruption* of an unreflective life, a *test*, or a *rebuke*.[4]

The authors of the *Malleus Maleficarum* of 1486, citing Peter Lombard, summarize the five causes of ailment in terms of the five ways that God addresses humanity:

For five causes God scourges man in this life, or inflicts punishment. First, that God may be glorified; and this is when some punishment or affliction is miraculously removed . . . Secondly, if the first cause is absent, it is sent that merit may be acquired through the exercise of patience, and also that inner hidden virtue may be made manifest to others. . . . Thirdly that virtue may be preserved through the humiliation of castigation . . . Fourthly, that eternal damnation should begin in this life, that it might be in some way shown what will be suffered in hell . . . Fifthly, that man may be purified, by the expulsion and obliteration of his guilt through scourges.[5]

James Rex, later King James I, simplifies their classification into two main categories, to *punish* and to *correct*:

God by the contrarie, drawes euer out of that euill glorie to himselfe, either by the wracke of the wicked in his justice, or by the tryall of the patient, and amendment of the faithfull, being wakened vp with that rod of correction.[6]

To say that all disease is a form of divine address is not to say that all disease is caused by God or even "willed by God." For there is a difference between God causing a disease, such as Nebuchadnezzar's madness, and God allowing disease to happen, as in Job. Still, whether by causing disease or by merely allowing disease to happen, God is weaving all things together for the greatest good, for God would neither cause nor allow disease unless it were to the increase of his glory. Thus King James boldly utters what might appear, at first glance, as a bit of shocking irreverence when he calls the devil "God's hang-man."[7] The joke, of course, is on the devil, who does not *know* he is God's hangman, who does not understand that even his machinations, too, have a role to play in the great drama called *the salvation of the world*.

But what is this *Christianization*? What is the decisive shift that transpired from the classical age to the Middle Ages within the purview of disease? The shift can be characterized succinctly: ambivalence versus concern. The Greek gods were, at best, *ambivalent* about humanity. This is why it was possible for Epicurus to entertain the thought that we need not worry about what lies beyond the grave: the gods live in a state of blessed unconcern about us. And while the followers of Asclepius, patron saint of medicine, praised him for his great love of humanity, he appeared to be the exception, rather than the rule: "If any god was interested in the private needs of men, in their most personal affairs, if any god showed providence, it was Asclepius."[8]

Now, however, we have a God who cares, who is focused and engaged, who is attentive to the welfare of the individual and, above all, to her salvation and everything that pertains to it: the texture of her everyday life, her spiritual longings, her carnal desires, her errant thoughts, her wayward glances, her moral lapses. Imagine taking a seat in a vast movie theater, the old-fashioned kind with the lush red curtain concealing the screen. The lights dim; the curtains part; silence ensues; we expect to be entertained by the antics of some foolish Greek gods. Yet on the other side, there is no screen. There is, instead, another room behind the curtain, containing a production set. There are bright lights shining on you that are nearly blinding. You see a camera pointed toward you, and behind all of those things, you discern the outline of a man, sitting in a chair, facing you. And you realize at that moment that *you* are the show, *you* are the spectacle, *you* are the evening's entertainment! And the story that you are in is a drama about salvation. Will you find the narrow gate? Or will you go the broad way that leads to destruction? Will you turn and receive the gift of grace through faith? Or will you remain in that sorry state of unreflection, desperately seeking ways to *kill time*?

Nor is God aloof, monitoring but impassive, like a psychotherapist sitting by the couch, occasionally nodding or frowning, and always writing everything in his book. From the beginning, he has been actively shaping the course of the narrative, gently nudging it this way and that, occasionally even entering the flow of history in the form of a man, first to warn, ultimately to judge. Yet, remarkably enough, he acts in such a manner that we, the actors, retain our power of choice, free will: nothing imposed; nothing forced.

The basic framework within which we must now understand madness is this newfound set of relationships between death, God, and the minutiae, the hidden folds, that make up an individual's life and that lead to its defining moment: will she choose redemption or perdition? All aspects of sickness and health must be rethought in light of this unprecedented structure. Whatever ails me, there is a reason for it, and while we may not know the immediate reason, we know the ultimate reason: the disease itself is the challenge or temptation that propels the narrative of my life forward, and brings the protagonist closer to the moment of decision, toward redemption or toward damnation, and, *if* redemption, that carries her along in the glorious movement of sanctification.

II.

It is in the context of this immovable and inviolable teleology of the world that we must understand demonology.

At the level of metaphysics, all diseases are from God: that much is settled. At this level, there is no coherent distinction between supernatural and natural causes of disease; all diseases are necessarily both, as even our Hippocratic author of *On the Sacred Disease* knew. But there is a practical level, a level of intervention, of agonized bodies and bloodletting, one at which we can and must distinguish the two. Some diseases, those caused by the *demonic*, require, perhaps in addition to natural remedies, direct appeal to God (*this kind can come out by nothing but prayer and fasting*).[9] So, within this Christianized world, *inside madness-as-strategy*, we can, practically speaking, reconstruct two classes of malady, one that falls under the jurisdiction of the physician and another that falls under the jurisdiction of the priest. One could, with some intellectual rigor, call the former disease "natural" and the latter "supernatural," but only if one bears continually in mind that the distinction pertains only to the immediate causes of the malady, that is, suffocation of the mother *versus* demon possession, and not to its ultimate cause, which is God's perfect will.

This pragmatic division between "natural" and "supernatural" forms of madness is recognized by both sides, the priests and the physicians: there is no question of the priests trying to appropriate the role of the physician, or vice versa. *On the one hand*, the priests acknowledge a boundary between natural and supernatural maladies. While *all* disease is divine strategy, *some* disease falls under the special jurisdiction of physicians, particularly that due to ill humor (accumulation or corruption of blood, choler . . .). The priests and exorcists have no desire to fold all madness under the banner of demon possession, or claim that "all madmen are afflicted by the demonic."[10] For demon possession, when placed alongside frenzy, mania, and melancholia, has a distinctive phenomenology. Marks of demon possession include speaking in strange languages, blaspheming God, exhibiting extraordinary physical strength, falling into convulsions and fits, even soothsaying. Demon possession is far different, "clinically speaking"—as we would now say—from the madness that stems from, for example, an abundance or corruption of blood or choler. Witness James:

There hath indeede bene an old opinion of such like thinges; For by the Greekes they were called λυκανθρωποι which signifieth men-woolfes. But to tell you simplie my opinion in this, if anie such thing hath bene, I take it to haue proceeded but of a naturall super-abundance of Melancholie, which as wee reade, that it hath made some thinke themselues Pitchers, and some horses, and some one kinde of beast or other.[11]

Burton, in his masterpiece, goes so far as to catalog demon possession as one *subtype* of madness, alongside natural types such as lycanthropia, hydrophobia, and St. Vitus dance.[12]

On the other hand, even physicians like Jorden did not deny that some forms of madness result from demon possession or witchcraft (which are, at root, one: the *Malleus Maleficarum* informs us that "in order to bring about evil a witch can and does co-operate with the devil").[13] After all, what constituted Jesus' earthly ministry but the triad of healing, preaching, and casting out demons? Are we to condemn Jesus, too? *This kind*, he tells his disciples, referring to a boy with seizures, *can only come out by fasting and prayer*. Furthermore, why does God, in Deuteronomy, forbid sorcery and witchcraft, if sorcerers and witches do not exist?[14] Convenient as it would have been for the physicians, none of them could claim, under pain of suspicion of heresy, that witches and demons are not real. Hence Jorden, despite his mocking tone, is quick to temper his condemnation of the exorcists:

I doe not deny but that God doth in these dayes worke extraordinarily, for the deliverance of his children, and for other endes best knowne unto himself; and that among other, there may be both possessions by the Divell, and obsessions and witchcraft, &c. and dispossession also through the Prayers and Supplications of his Servants, which is the onely meanes left unto us for our reliefe in that case. But such examples being verie rare now adayes, I would in the feare of God advise men to be very circumspect in pronouncing of a possession: both because the impostures be many, and the effects of naturall diseases be strange to such as have not looked throughly into them.[15]

Jorden's aim is neither to eliminate the role of the supernatural in disease, nor the place of the priest in healing, but to establish the *proper jurisdiction* of physicians. It is a question of firmly insisting that, at least in principle, *we, the physicians, have jurisdiction over some forms of madness, and you, the priests, have jurisdiction over others.* Ultimately, Jorden will define the former as those that involve organic dysfunctions. In this way, the study of madness is gently unmoored from the divine.

We should think of what Jorden is doing, then, as finding a safe haven for the physician, *inside* of an explicitly theistic setting. It is not a question of rejecting or challenging God's sovereignty over disease but demonstrating that his sovereignty itself allows for, and coexists with, the labor of the physician, within certain well-defined boundaries. And there are various strategies that Jorden uses to carve out this space, to construct this haven for allowing the physicians to practice medicine as they see fit. First, he argues that, even of those diseases that are demon-caused, which are many, there are two very different ways in which demons cause sickness. In some cases, demons cause sickness in the "canonical" way, through possession, by using the human body as a host. Such cases fall squarely under the jurisdiction of the exorcist. In other cases, however, the demon causes disease *indirectly*; the demon disrupts the course of nature by, say, inducing an excess or corruption of one of the humors, and then *fleeing*. The *physician* has jurisdiction over those cases:

> If the divell as an externall cause, may inflict a disease by stirring up or kindling the humors of our bodies, and then depart without supplying continuall supernaturall power unto it; then the disease is but naturall, and will submit it selfe unto Physicall cure. For externall causes when they are already remoted, give no indication of any remedy.[16]

In short, if the demon causes the disease merely by tampering with the proportion or integrity of the humors, then the proper course of treatment is to return the four humors to their proper balance, and *that* task falls to the physician, not the priest.

But—and this is his second strategy—not only does the physician have jurisdiction over those cases that do not strictly involve possession, the physician has jurisdiction over *how to draw the line between those cases*. That is, the physician alone has the special authority to decide, of any particular case, whether it is strictly a case of possession. *The physician has meta-jurisdiction*, jurisdiction over the question of jurisdiction:

> But the learned Phisition who hath first beene trained up in the study of Philosophy, and afterwards confirmed by the practice and experience of all

manner of naturall diseases, is best able to discerne what is naturall, what not naturall, what preternaturall, and what supernaturall, the three first being properly subject to his profession.[17]

A third strategy is alluded to earlier ("I doe not deny but that God doth in these dayes worke extraordinarily . . .") but not developed systematically in his book.[18] One line some Protestant preachers, such as John Walker and John Deacon, took up is that while there *is* a role for the exorcist, the role of the exorcist has largely, if not entirely, faded from history. It is true that a significant part of Jesus' earthly ministry was casting out demons and performing other miracles the likes of which are rarely seen today. *But the era of miracles is over.* The reason that faith healing and exorcism played such a significant role in the Apostolic Age was that God was in the process of establishing his church. The extraordinary acts of healing, of exorcism, even of raising the dead, performed an *evidentiary and authenticating function* ("but that you may know that the Son of Man has authority on earth to forgive sins . . . ").[19] Miracles had the function of validating the message of Jesus and his followers and thereby establishing his church. But now that the church and holy scripture are here, miracles are no longer necessary. Holy scripture is *self-authenticating*.

III.

By identifying the proper jurisdiction of the physician, Jorden sought to wrest a measure of control of madness from the hands of the exorcists. What, then, *is* the physician's proper jurisdiction? Over what realm does his authority command absolute respect? The physician has unconditional authority over the realm of *organic dysfunction*, that is, the inability of the parts of the body to perform their God-given duties and roles, regardless of whether the cause of that dysfunction is physical, such as excessive heat, mental, such as anxiety or fear, or spiritual, such as demonic tampering. His authority ends when he comes up against strict possession, but even then, he has the final authority to make that determination.

To this end, he advocated for the recognition of a relatively new kind of disease, the *suffocation of the mother*, or hysteria. While Jorden did not invent the disease, he brought it into a new prominence in England. Many cases like Glover's at the time would have been classified as melancholy, not hysteria.[20] What is remarkable about the *mother* is its explanatory promiscuity; it can account for *any and all* forms of women's suffering, particularly as it manifests itself in the mental sphere. That is because the mother, that is, the womb, has the curious power to *mimic* diseases that would typically have fallen under the priest's jurisdiction. The mother can give rise to "*suffocation* in the throate, chowing of Cockes, barking of Dogges, garring of Crowes, frenzies, convulsions, hickcockes, laughing, singing, weeping, crying, &c."[21]

What is this extraordinary disease? And what lends it this exceptional power of mimicry? Its power proceeds from its unique placement in the woman's body. Because of its pivotal role in generation, the womb is served by all of the major organ systems and, in turn, serves all of the others. It has rich interconnections to the brain, seat of the animal functions; to the heart, seat of the vital functions; and to the liver, seat of the natural functions. Disturbances, therefore, in the mother, can affect any and all of these diverse organs. And the disturbances that it can induce are not merely those that originate with a dysfunction or breakdown "inside" the mother. A dysfunction or breakdown *anywhere* in the body may have a pathological impact on the mother *and thereby* communicate itself to almost any other part of the body. It both initiates and transmits pathology. Consider the case of a "venomous vapour, arising from this corrupt humor unto divers parts of the bodie."[22] The mother, to the end of self-preservation, constricts itself and begins to rise into the body, where it wanders in search of a safe haven. Forthwith—depending on its precise trajectory—it can press into any organ and create new disturbances: "for that the matrix being grievously anoyed with the malignity of those vapours doth contract it selfe and rise up by a local motion towards the midriff."[23]

As in Hippocrates, Jorden depicts the different symptoms of hysteria as a result of so many different ways that the body's natural faculties can fail. This is his version of "differential diagnosis." He begins with a sketch of the functional organization of the human body: the vital, the animal, the natural, and he specifies the function of each one. What is extraordinary about the mother is its ability to *diminish, deprave, or abolish* the function of any one of those faculties in highly specific ways.[24] For example, consider the way the animal faculty, or any of its sub-faculties—sensation, intellect, movement—can be disrupted. A disruption of the faculty of *sensation* will lead to the seeming demoniac's well-known tolerance, even complete unawareness, of the pain of pinpricks and burning. A disruption of the faculty of *movement* leads to choking. And so on, for all of the symptoms associated with possession. Suffocation, convulsions, noises in the throat, trembling, laughing, weeping—all can be traced to a breakdown of specific functions.

In fact, because of its ubiquity, because of its explanatory flexibility, we should not even think of suffocation of the mother as a specific "disease entity," the way that we think of abundance of melancholy, or even *Bacillus anthracis*. It is not one among other natural causes, *but natural cause itself*. It is the form of natural cause; it is one with the process of pathogenesis. To explain a certain manifestation, say, croaking like a frog, as due to suffocation of the mother, is simply to affirm the presence of an organic dysfunction, and to insist that the symptomatology falls under the jurisdiction of the physician. The mother is a form without content, a *template* for disease explanation; it is vacuity incarnate. It is the pinnacle of what is now disparaged as "medicalization." Today we often lament the "medicalization" of this or that—the medicalization of sadness, of anxiety, of the social misfit. Whatever this medicalization turns out to be, we should see Jorden's appeal to *suffocation of the mother* as a kind of ideal representative of medicalization itself. We can take quite literally any mental disturbance we would like, *if in a woman*, and say of it, "This is a medical condition; it is due to the mother."

The mother is disruption itself. Her job is to disrupt; she is an agent of dysteleology. And hence, we can see Jorden, in his rebuke of the exorcists, not only as carving out a space for the physician but as carving out a space for the unimpeded exercise of dysteleology, a reign of dysteleology that need not be perpetually subordinated to God's ordered plan, need not be constantly referred to this plan, illuminated in terms of the plan. We witness the ascendency of madness-as-dysfunction *inside* madness-as-strategy, like a virus proliferating within its host.

Jorden is famous not only for the suffocation of the mother, but for acting as an expert witness in the case of Mary Glover. In 1602, Elizabeth Jackson was indicted for witchcraft for afflicting Mary Glover with terrible ailments. Jorden, testifying on behalf of Jackson, argued that these afflictions did, in fact, have a natural cause, and that Jackson was therefore not a witch. Hence, the case of Mary Glover was one of the events that brought the tension between physician and priest into the public eye and partly stimulated the writing of his book. In a contemporary context, it would be easy to see Jorden as a hero of science, rationalism, and, in some measure, women's rights.

In this case however, the principal judge, Sir Edmund Anderson, ultimately decided against Jackson; Mary Glover, in fact, was demon possessed. He rejected Jorden's medical testimony, and *rightly so*, recognizing the vacuity of the medical explanation Jorden offered: "Then in my conscience, it is not naturall; for if you tell me neither a Naturall cause of it, nor a naturall remedy, I will tell you, that it is not naturall."[25] His reasoning was correct: Jorden did not state a cause for Glover's ailments but offered the vacuous form of natural cause.

Even in his manuscript, Jorden admits that it is very difficult to explain the inner dynamics of the mother in vulgar tongue, and therefore that he would refrain from saying too much about the mechanics of causation:

> The causes of this disease and of the *Symptoms* belonging therunto, have ever bin found hard to be described particularly: and especially in a vulgar tongue, I hold it not meete to discourse to freely of such matters, and therefore I doe crave pardon if I do but slenderly overpasse some poynts which might be otherwise more largely stood upon.[26]

His rhetoric suggests, of course, that he has many, rigorously documented, things to say about the mother, that he possesses a wealth of scientific information, and it has been written in other places, for other professionals, and nobody should expect everything to be said all at once, particularly not in a small book such as this, intended for popular consumption, et cetera, et cetera.

IV.

The irony of Jorden's attempt to make a space for madness-as-dysfunction inside madness-as-strategy is that, even for those cases of madness that fall under the jurisdiction of the physician, he does not entirely *banish* the priest, the exorcist, the witch-hunter, for its treatment. Instead, *he finds a way to incorporate them, without any perceived contradiction, into his arsenal of treatment.* It is not enough to expel the purifications, the charms, and the exorcisms. We can include them, too, in our inventory of remedies, once we have divested them of their supernatural power. The charms, incantations, purifications, have been robbed of the supernatural and redeployed as theater. What allows for this—what *requires* it—is his recognition that some distemper can be caused by perturbations of the mind. The experience of intense emotions such as love, joy, shame, anger, and fear, can change the body in such a way as to induce disease. A wise physician must, then, be prepared to engage with the patient not only at the level of nature but at the level of mind. And to engage mind, we need both the tools of reason (counsel, religion, etc.) *and* the tools of unreason (charms, exorcisms, etc.):

> So that if we cannot moderate these perturbations of the minde, by reason and perswasions, or by alluring their mindes another way, we may politikely confirme them in their fantasies, that wee may the better fasten some cure upon them. . . . The like opinion is to bee helde of all those superstitious remedies which have crept into our profession, of Charmes, Exorcismes, Constellations, Characters, Periapts, Amulets, Incense, Holie water . . .[27]

Again, Jorden is affirming not the supernatural efficacy of those unreasonable tools, but their psychological efficacy: we employ them because they can stir the passions in the right way. In this connection, it should not be surprising that he approvingly cites Galen, who

> boasteth that he did every yeare cure many diseases by this stratagem of moderating the perturbations of the mind by the example of *Aesculapius* who devised many songs and ridiculous pastimes for that purpose.[28]

The question is not one of banishing the priest, then, but of giving him a new job to perform. The priest can speak to the mad person, so long as we, the physicians, understand that he does not speak the truth of disease: his rituals, his purifications, his incantations, are a sham, a ploy, a play, *a way of defeating unreason through unreason itself*. We will see this trope recur as a constant refrain in nineteenth-century medical textbooks: to cure the madman's delusion, you must step into the delusion; you must play along with it; you must gradually undermine it from within.

Robert Burton, in his momentous *The Anatomy of Melancholy* of 1621, developed the dual teleology of madness. But he also began the process of mapping this teleological framework onto the order of nature. The question is not, for Burton, one of *replacing* madness-as-strategy with madness-as-dysfunction, but showing how God's plan for salvation is etched into the ironclad laws of nature. For Burton, diseases are not *merely* punishments, and they are not *merely* opportunities for redemption; they are also inexorable consequences of the misuse of our God-given faculties. In other words, God has designed the world in such a way that the misuse of our God-given faculties will, inexorably and naturally, induce madness. The requirement for punishment is embedded in the causal structure of reality. If you overindulge in food or drink, *you get sick*. This does not require a *special* intervention on God's part; it is a consequence of the way that our bodies are designed, but God designed our bodies to react that way, *so that we would learn not to overindulge*. This recognition—that the punishment for sin can be woven into the causal structure of the world—brings with it a further unmooring of the study of madness from theology, for when we encounter madness, we ask not, or not merely, *For which sin is God punishing you?*, but *What are the natural mechanisms that tend to ensure the causal connection between that particular sin and that particular form of madness?*

3

Madness as Misuse and Defect

Robert Burton, in his momentous *The Anatomy of Melancholy* of 1621, embraces this dual teleology of madness—madness always has the twin characteristics of punishment and opportunity for redemption—and these two characteristics of madness mirror two attributes of God, his *justice* and his *mercy*. Yet this requirement for punishment does not always necessitate a special intervention on God's part: rather, God has designed nature in such a way that certain kinds of sins will be followed by certain kinds of diseases, such as gluttony and indigestion (Section I). This suggests that madness must be seen, in Burton, not as *defect* but as *misuse*, that is, as a just consequence of the rebellious misuse of our God-given faculties. Furthermore, Burton shows precisely how the forms of madness are not just punishments, but *portraits*, of our sin. The purpose of any particular form of madness is to expose the nature of the underlying sin that God wants us to repent of (Section II). In some ways, it is more illuminating to read *The Anatomy of Melancholy* as a morality tale, on the order of *The Divine Comedy* or the "Macro Plays," rather than as a medical treatise (Section III).

I.

God is the god of justice *and* of mercy. His justice is complete; it is exact; it is surgical in its precision. When Nadab and Abihu recklessly disobeyed God by bringing unauthorized incense into the temple, they were immediately consumed by fire. This might appear to us as a harsh penalty or as disproportionate to the crime; it might appear to the unredeemed mind as sheer cruelty. But that is because we can scarcely understand—we rarely experience—God's justice in its purity, its immediacy, its perfection. Justice flows out of God's nature like an unstoppable river.

God is also the god of mercy. God loves all people and wants them to be saved. The story of Jesus and the adulteress, though it might be an interpolation, accurately captures how God thinks. The men who picked up stones

Madness. Justin Garson, Oxford University Press. © Oxford University Press 2022.
DOI: 10.1093/oso/9780197613832.003.0004

to hurl at the adulteress were technically and legally in the right; they were carrying out the God-ordained punishment for that crime, *and Jesus never rebukes them*—but he encourages them to relent nonetheless.

But the combination of these two characteristics, God's justice and God's mercy, evokes a paradox, if not a contradiction: how can God be, totally and completely, the god of justice *and also*, totally and completely, the god of mercy? For, to the extent that God is the god of justice, God will punish wrongdoing completely, perfectly, and without remainder. As the god of mercy, God will forgive, and forgiveness is nothing if not releasing someone from the rightful claims of justice, allowing them to forgo a deserved punishment. So, mercy seems to demand injustice, and justice, to be justice, must be carried out without a shred of mercy. Both seem to demand from God something contrary to his being.

In Christianity, of course, these two aspects of God's character, his justice and his mercy, are reconciled in the crucifixion and resurrection of the Christ. For in Jesus, God takes the sin of the world and imputes it *to his own flesh*, where he punishes it totally and completely. This is why Paul can say that Jesus is the propitiation (ἱλαστήριον, having an expiating or placating power), for our sins, and this is why the author of Hebrews can see, in the crucifixion, a portrait of the atoning sacrifice of the lamb required in Leviticus: unlike the sacrifice of a lamb or goat, which must be carried out annually as substitutionary atonement, the crucifixion is the ultimate and final sacrifice, "for all time a single sacrifice for sins."[1] That God must allow his son to be killed is neither cruelty nor capriciousness, as some have perversely set forth. It is a necessary consequence of his moral perfection. So, when God, in his mercy, admits a sinner like me into his kingdom, it is *not* as if the claims of justice have been imperfectly satisfied. It is not as if, in his kindness, he has chosen to forgo the penalty that I deserved. God could not do that and still be God. God doled out that penalty at Calvary and I accepted it. Nobody can claim to understand Christianity who does not grasp how the crucifixion and resurrection carry out a seemingly impossible synthesis of justice and mercy, that is, unless they understand how the cross presents a solution to a seeming tension intrinsic to the being of God.

In the medieval imagination, madness, too, represents a divine synthesis of justice and mercy. Madness is always and truly *by design*. It embodies both aspects of God's character: this is the *dual teleology* of madness. On the one hand, madness always fulfills God's justice; it is a punishment for sin, for turning away from or rejecting God's wise counsel. On the other hand, the form that the punishment of madness takes is designed to make us aware of our wrongdoing, and thereby create an opportunity for redemption. Burton, right at the outset of the first partition of his book, announces both aspects of God's character, starting with the punitive:

> The impulsive cause of these miseries in man, this privation or destruc-
> tion of God's image, the cause of death and diseases, of all temporal and
> eternal punishments, was the sin of our first parent Adam, in eating of the
> forbidden fruit, by the devil's instigation and allurement. His disobedi-
> ence, pride, ambition, intemperance, incredulity, curiosity; from whence
> proceeded original sin and that general corruption of mankind, as from
> a fountain flowed all bad inclinations and actual transgressions, which
> cause our several calamities inflicted upon us for our sins . . . God is angry,
> punisheth and threateneth, because of their obstinacy and stubbornness,
> they will not turn unto Him . . . To punish therefore this blindness and ob-
> stinacy of ours as a concomitant cause and principal agent, is God's just
> judgment in bringing these calamities upon is, to chastise us, I say, for our
> sins, and to satisfy God's wrath.[2]

Madness is simply *one malady among others* that God uses to satisfy his just wrath: "The LORD will afflict you with madness, blindness and confusion of mind."[3]

We have already encountered, in the ancient magicians, the idea that madness could serve as a punishment, that is, how madness traces its teleological character back to its ultimate source, namely, an offense against the gods and a divine punishment for that offense. By the Middle Ages and the Renaissance, as we said, this basic understanding is Christianized; both the physicians and the priests understood that, in addition to its character as divine punishment, madness—and disease generally—also embodies God's mercy. But in what respect, precisely, does madness announce God's mercy? For Burton, the merciful character of madness is seen in the fact that madness, and disease generally, gives the sufferer an opportunity to reflect on his wayward path and thereby seek redemption:

> Or else these chastisements are inflicted upon us for our humiliation, to
> exercise and try our patience here in this life, to bring us home, to make us
> know God ourselves, to inform and teach us wisdom . . . He is desirous of
> our salvation . . . and for that cause pulls us by the ear many times, to put us

in mind of our duties: "That they which erred might have understanding" (as Isaiah speaks, xxix, 24), "and so to be reformed." . . . *In morbo recolligit se animus*, as Pliny well perceived; "In sickness the mind reflects upon itself, with judgment surveys itself, and abhors its former courses."[4]

In this respect, these passages from Burton merely summarize the reigning consensus on madness; they add nothing new to it.

But with Burton we see a new structure emerge, a new way of thinking through this dual teleology of madness. Madness, insofar as it is a punishment for sin, no longer requires God's direct, supernatural intervention into the course of nature. Rather, God has woven the punitive requirement into the causal order of reality. Punishment no longer, or no longer *always*, takes the form exemplified by Nadab and Abihu, who were struck dead instantly for bringing unauthorized incense into the temple, or even the form exemplified by Nebuchadnezzar, who was immediately struck with madness because of his unspeakable hubris. Rather, the penalty of madness is more akin to the penalty that attaches to gluttony or drunkenness. *Given the causal structure of reality*, that is, given the way that the human frame is designed, intemperance will lead to circulation problems, to organ failure, to unnecessary bodily pains, to stroke, even to early death. Put crudely, once God arranged the causal structure of reality, he did not "need to do anything else," as the philosophers say, to ensure that certain actions would be attended by certain punishments. So, even though with Burton we see something of a "naturalistic" turn in thinking about madness, it is *not* naturalistic in the sense that it departs from a fundamentally theistic framework: it never stops seeing, in madness, a marriage of justice and mercy. It is naturalistic only in the sense that it sees, in the punitive aspect of madness, a regular, law-like, nomological character, and it can infer this character from the very design of the human body.

This is why it would be anachronistic, from a historical point of view, to pose the question "Does Burton adopt a *natural* or a *supernatural* approach to medicine?" It would be mistaken to say that Burton takes a natural, non-supernatural approach; it would also be mistaken to say that he takes a supernatural, non-natural approach; *and it would be equally mistaken to say that he takes* both *a natural and a supernatural approach*: to say that he takes both is, potentially, to treat these, a supernatural and a natural approach, as if they *could be* separated, as if we *could* attribute independent causal agency to each. This is an exact parallel to the problem, in the philosophy of biology, of *apportioning causal responsibility* to genes or environment.[5] In the Renaissance, the categories of supernatural and natural interpenetrate in a way that is difficult for us moderns, us denizens of Taylor's secular age, to come to terms with. One cannot even begin to understand the causal structure of reality without understanding God's plan for humanity. Understanding God's character is not merely a benefit of the study of nature; it is a prerequisite, too. Burton articulates the sheer impossibility of separating the two, and even the interpenetration of both, when he explains why he chose, of all the diseases one could study, melancholy, this *compound mixed malady*:

> It is a disease of the soul on which I am to treat, and as much appertaining to a divine as to a physician, and who knows not what an agreement there is betwixt these two professions? . . . Now this being a common infirmity of body and soul, and such a one that hath as much need of a spiritual as a corporal cure, I could not find a fitter task to busy myself about, a more apposite theme, so necessary, so commodious, and generally concerning all sorts of men, that should so equally participate of both, and require a whole physician. A divine in this compound mixed malady can do little alone, a physician in some kinds of melancholy much less, both make an absolute cure.[6]

Jorden, if you recall, was preoccupied, above all else, with a *demarcation* problem: *in which circumstances* do you need a physician, and *in which circumstances* do you need a priest? His goal was to marginalize the role of the priest as much as possible *short of* denying entirely the existence of the supernatural. *If* demons cause madness merely by stirring up ill humors in the body, and departing forthwith, then the cure of madness falls exclusively under the jurisdiction of the physician, not the priest, because, after all, treating the patient is simply a matter of restoring the humors to their proper proportions. From Burton's point of view, this reasoning is not merely mistaken, but asinine: if madness is a *natural punishment* for sin, then its treatment requires both the physician and the priest. After all, what would be the lasting benefit of restoring the humors to their proper proportions if one does not eradicate the same moral cause, the same character flaw, that produces madness?

That is why, though Burton divides madness into two main classes, *supernatural* and *natural*, at the same time he recognizes that this distinction is not a proper distinction, not a distinction in *sensu stricto*. For he insists of *all* disease that it has this twofold purpose of punishment and redemption:

> To punish therefore this blindness and obstinacy of ours as a concomitant cause and principal agent, is God's just judgment in bringing these calamities upon us, to chastise us, I say, for our sins, and to satisfy God's wrath.[7]

This statement is intended to apply to all diseases, including madness, and not just the nominally "supernatural" ones.

What, then, makes a disease *supernatural* rather than *natural*? That is, to the extent that he recognizes some kind of distinction, what is that distinction? We can say, provisionally, that for Burton, a disease is "supernatural" to the extent that we can see in it a kind of immediacy, a *fittingness* (in a sense to be described), and the relative absence of obvious physiological triggers: these are the diseases that come immediately from God or from demons, or mediately from witches and magicians. Such diseases, and their variations, bear the unmistakable stamp of the spiritual. When Nebuchadnezzar equated himself to God, he was immediately struck with madness that lasted for seven years. The punishment was swift, it was fitting—in a sense to be described—and there were no obvious natural causes. In his later years, Linnaeus assembled a massive number of such examples in an unpublished manuscript, under the title *Nemesis divina*, with the same fervor with which he once cataloged plants. He sought to demonstrate that the just wrath of God was evident to anyone with eyes to see.

What, then, makes a disease natural? At the outset, and again, in a highly provisional way, Burton says that a disease is natural when God uses nature as instrumental cause in satisfying his purposes:

> Now the instrumental causes of these our infirmities are as diverse as the infirmities themselves; stars, heavens, elements, etc., and all those creatures which God hath made, are armed against sinners. They were indeed once good in themselves, and that they are now many of them pernicious unto us, is not in their nature, but our corruption, which hath caused it.[8]

A natural disease, however, cannot simply be *identified with* one that is caused by God through the intermediary of nature. For in that case, we would have to classify the flood of Noah and the plague of serpents as "natural," but Burton classes those as supernatural.[9] Rather, we should say that a disease is natural when there is something of a law-like necessity that ties the sin to the punishment—for example, gluttony and indigestion, or heavy drinking and liver damage.

But if what makes a disease *natural* is that there is a law-like necessity that ties the sin to the punishment, then natural disease is not something that God, as it were, inflicts on us, as with the plague of serpents. It is something that *we inflict on ourselves.* Our sin, in conjunction with the God-given laws of nature, produces the disease. Specifically, "natural" disease represents the law-like outcome of the willful *misuse* of our God-given faculties, as we will see.

Another way of understanding the notion of a "natural" cause of madness is through Paul's famous admonition that *you reap what you sow*: "For he that soweth to his flesh shall of the flesh reap corruption; but he that soweth to the Spirit shall of the Spirit reap life everlasting."[10] The proverb does not merely express the principle of just desert; it does so by deploying an agricultural metaphor. The image centers around planting a seed, watering the ground, *letting nature take its course*, and then, at long last, judging the result. God's moral vision is wedded to the rhythms of nature. If I sow the seeds of animosity, anger, lust, covetousness, I ultimately reap the penalty for that; if I sow the seeds of kindness, gentleness, patience, I ultimately reap the fruits of the Holy Spirit. Another example of this principle is found in the parable of the wheat and the tares. A good seed, working in tandem with natural processes, will yield wheat; a bad seed, working in tandem with natural processes, will yield weeds. At the harvest, which is always the day of judgment, the moral character of our actions will be manifest; they will yield their fruit: something beautiful and useful, or something hideous and useless, fit only to be gathered and burned.

II.

For Burton, the organizing principle for understanding madness is *not* the category of defect or dysfunction, but the category of *misuse.* The willful misuse of one's faculties works in tandem with natural processes to yield the fruit of madness. *Dysfunction* or *defect* is something that *befalls* one, something that one is passive in relation to. *Misuse* is active, intentional, willful; it is morally evaluable. This is why liability law makes a strict separation between those harms that are due to manufacturer's defects and those harms that are due to consumer misuse, that is, the use of the product in a manner or in an environment that it was not designed for. The idea of madness as a form of misuse *rather than* defect not only plays a central explanatory role in Burton's text, but also serves as a *theodicy*: there is no reason to blame God for madness, to shake one's fist at the heavens for one's ailments. As a rule—and setting aside hereditary madness and madness acquired in the course of nursing, which require a separate analysis—madness stems from the *freely chosen and willful misuse* of our God-given faculties. True, God's will is always involved in madness—all madness synthesizes God's justice and mercy—but the question here is one of blame, of seeking redress, of attributing wrongdoing. To the extent that it is a question of blame, the blame is squarely on the shoulders of the mad person, not on God.

An apt metaphor that Burton uses for understanding madness *as* resulting from misuse, rather than defect, is that of a perfect sword that a man uses to impale himself:

> We study many times to undo ourselves, abusing those good gifts which God hath bestowed upon us, health, wealth, strength, wit, learning, art, memory to our own destruction. . . . [W]e arm ourselves to our own overthrows; and use reason, art, judgment, all that should help us, as so many instruments to undo us. Hector gave Ajax a sword, which, so long as he fought against enemies, served for his help and defence; but after he began to hurt harmless creatures with it, turned to his own hurtless bowels. Those excellent means God hath bestowed on us, well employed, cannot but much avail us; but if otherwise perverted, they ruin and confound us: and so by reason of our indiscretion and weakness they commonly do, we have many instances.[11]

A man impaled on his own sword: this is a portrait of what happens when we misuse God's perfect faculties. Much of book 1, partition 1, of the *Anatomy* is simply a catalog of diverse mental maladies, and, corresponding to each, the distinctive vice that tends to induce it.

Madness stems from the misuse of our God-given faculties, and not from a dysfunction or defect. But we have not yet exhausted the structure of madness. As we have shown, Burton reconciles two qualities that sit in tension with each other: God's justice and God's mercy. So, we must see how each of these two aspects of God's character are realized in every form of madness. *As a rule*, we find in each form of madness two aspects.

Madness is a *natural consequence* of our sin, but it is also a *portrait* of our sin. The first aspect of madness corresponds to God's justice. The second aspect of madness corresponds to God's mercy.

Consider the first aspect of madness, madness as exemplifying God's justice. Madness is a natural consequence of the sinful misuse of our faculties. It is a punitive consequence. In other words, for God's justice to be realized, we must suffer in some measure for our sin. But the second, and equally important, aspect of madness, madness as exemplifying God's mercy, can be seen in this: madness is, as a rule, a *picture* of our sin. It is a portrait of our sin. More often than not, it is a caricature of our sin; it demonstrates our sin in an exaggerated or hyperbolic form. As a mirror of my sin, madness draws my attention to that area of sin inside of me, that region of rebelliousness that God wants me to repent of. In this respect, madness exemplifies God's mercy.[12]

We can see these two aspects of madness clearly when we turn back to Nebuchadnezzar. Nebuchadnezzar's sin was the sin of pride. When he looked upon all that he had accomplished, he said, "Is not this great Babylon, which I have built by my mighty power as a royal residence and for the glory of my majesty?"[13] Immediately forthwith, he was driven into the fields, "to eat grass like an ox." Here, the punishment fits the nature of the sin. In Nebuchadnezzar, we have an insignificant clod of dirt equating itself to God almighty; as punishment, God renders him an insignificant beast and forces him to degrade himself for seven years by eating grass with the other beasts. But in the form of the punishment, God created a *portrait*, or a caricature, of Nebuchadnezzar's sin: *you think you are equal to me, but this is what you look like in my eyes: a lowly beast eating grass from the field.* The crucial factor is that the specific form of the punishment was itself an act of mercy. *That* Nebuchadnezzar was to be punished is required by justice. That, *in* the punishment, Nebuchadnezzar was given a portrait or caricature of his sin—in that, God shows mercy. God revealed to Nebuchadnezzar the nature of the sin in his heart that he wanted Nebuchadnezzar to repent of. And repent he did, for after seven years, after confronting the nature of his sin, he was restored to sanity to praise his maker: "At the end of that time, I, Nebuchadnezzar, raised my eyes toward heaven, and my sanity was restored. Then I praised the Most High; I honored and glorified him who lives forever."[14] He was healed. The disease of madness healed the disease of hubris.

Again, as a rule—and excluding hereditary diseases and those acquired in infancy—to the extent that madness is a natural consequence of the misuse of our God-given faculties, it embodies God's justice; to the extent that madness presents us with a picture or portrait of our sin, it embodies God's mercy.

Burton playfully introduces the theme of madness as a *portrait* of sin in one of the two poems that open the book, a poem that describes the book's frontispiece. The lower right corner of that frontispiece shows a portrait of *Maniacus*, the disease of madness *sensu stricto*. Madness, in the narrow sense, differs from melancholia by being accompanied by fits of uncontrollable rage. Here, Burton colorfully describes the maniac as a *portrait* of unchecked anger:

> But see the Madman rage downright
> With furious looks, a ghastly sight.
> Naked in chains bound doth he lie.
> And roars amain, he knows not why.
> Observe him; for as in a glass,
> Thine angry portraiture it was.
> His picture keep still in thy presence;
> 'Twixt him and thee there's no difference.[15]

This form of madness has its source in unchecked anger, which is a sin, the sin of unforgiveness. Suppose I am humiliated—say, I have been demoted at work for some alleged breach of conduct. I am angry at my superiors; I am angry at the institution; I think of nothing but revenge. Suppose I *stew* in my anger; I ruminate on it; I fantasize constantly of retaliation. I could have accepted the Savior's sage advice to forgive my enemies as God has forgiven me. I could have accepted the wisdom of Paul not to let the sun go down on my anger. I could have taken solace in the knowledge that God will judge eventually. By ruminating on my anger, I willfully reject that advice. That constitutes a sin. And the natural consequence of this sinful rumination is maniacal fury.

But how does the sin of unforgiveness become converted into maniacal fury? Through a simple core psychophysiological mechanism: *disposition becomes habit*. The importance of this mechanism, as Radden has recently argued, cannot be overstated.[16] When we indulge repeatedly in a certain pattern of vice—anger, lust, covetousness—we change our constitution, we pervert our temperature, we stir up choler; that pattern of vice becomes a fixed habit, *and that habit, so fixed, is itself a form of madness*:

"Anger and calumny (saith he [Lucien]) trouble them at first, and after a
while break out into madness: many things cause fury in women, especially
if they love or hate overmuch, or envy, be much grieved or angry; these
things by little and little lead them on to this malady." From a disposition
they proceed to an habit, for there is no difference between a madman and
an angry man in the time of his fit.[17]

Disposition becomes habit: this is why madness, *sensu stricto, must* resemble
anger, must be a "portraiture" of anger, because *madness is just crystallized
anger*. When Burton writes, "Observe him [the madman]; for as in a glass,
thine angry portraiture it was," he is not using poetic license; he is not engaging in hyperbole; he is enunciating his principled conviction that the only
difference between a sane person in a fit of anger and the fury of a maniac is
the fixity and duration of his anger, not its quality:

> They [the angry men] are void of reason, inexorable, blind, like beasts and
> monsters for the time, say and do they know not what, curse, swear, rail,
> fight, and what not? How can a madman do more? . . . If these fits be im-
> moderate, continue long, or be frequent, without doubt they provoke
> madness.[18]

The same mechanism—disposition becomes habit—is at work in melancholy. For melancholy is but the result of the continuous indulgence in grief
over the ordinary tragedies of life. The difference between a melancholic and
a sane person is *not* that the one experiences tragedy and the other does not;
life is terrible for everyone:

> In a word, the world itself is a maze, a labyrinth of errors, a desert, a wil-
> derness, a den of thieves, cheaters, etc., full of filthy puddles, horrid rocks,
> precipitiums, an ocean of adversity, an heavy yoke, wherein infirmities
> and calamities overtake and follow one another, as the sea-waves . . . and
> you may as soon separate weight from lead, heat from fire, moistness from
> water, brightness from the sun, as misery, discontent, care, calamity, danger,
> from a man.[19]

The difference between the sane person and the melancholic is that
the latter indulges in his grief until it stirs up the black bile, perverts his

temperature, and becomes a more or less fixed characteristic. Through this, the grief of melancholy becomes a portrait, and even a caricature, of ordinary sadness:

> So most especially it [the imagination] rageth in melancholy persons, in keeping the species of objects so long, mistaking, amplifying them by continual and strong meditation, until at length it produceth in some parties real effects, causeth this and many other maladies.[20]

We see once more this dual teleological structure in covetousness, which leads to paranoia. Suppose I am a greedy, stingy man. With time, my disposition toward greed becomes a fixed habit; I start to distrust my own family; I believe that people are coming for my wealth; I start to hear of conspiracies against me. The paranoiac is simply a caricature of covetousness; it is what covetousness looks like in its inmost nature: "They are in continual suspicion, fear, and distrust. He thinks his own wife and children are so many thieves, and go about to cozen him, his servants are all false."[21]

This is not to say that every form of madness in the *Anatomy* exhibits a precise mirroring function. Not every kind of madness is a portrait of a corresponding vice. The point here is not to impose a procrustean formula for analyzing Burton, but to draw out a theme or thread that runs throughout his complex vision. But even when the form of madness is *not* a portrait of the corresponding sin, we can still discern a kind of divine *fittingness* between the madness and its corresponding vice. (This is not a *fittingness* in the sense that the punishment must fit the crime; all punishment must be fitting in *that* sense in order to be punishment; there is no question that God's punishments are fitting in this trivial sense.) Rather, in all cases, the form of madness is fitting to the corresponding vice in the sense that the contours of the punishment *reveal the truth* of the corresponding sin.

Consider the melancholy that stems from immoderate gaming, drinking, or whoredom. The natural end of the unhinged pursuit of immoderate pleasure is melancholy; "as bitter as gall and wormwood is their last, grief of mind, madness itself."[22] All of these result in destitution:

> It is a wonder to see, how many poor, distressed, miserable wretches one shall meet almost in every path and street, begging for an alms, that have been well descended, and sometime in flourishing estate. . . . 'Tis the

common end of all sensual epicures and brutish prodigals, that are stupefied and carried away headlong with their several pleasures and lusts.[23]

Now, destitution is by no means an obvious *portrait* of immoderate indulgence in drinking, gaming, or whoredom. Nonetheless, the grief of destitution is a *fitting end* to those vices because it *reveals the truth* of the immoderate pursuit of pleasure: those pleasures are fleeting; they are hollow; they are empty; they are as nothing, and hence my being brought to nothing, to destitution, is an appropriate revelation of the truth of my journey. The prodigal son, at the height of his despair, looked longingly at the pods that were meant to feed the swine; his utter destitution reflected the emptiness of the pleasures he sought. We speak of "illusory" pleasures. The illusion is that they appear as if they will be satisfying and fulfilling, but they are hollow, disappointing. Hence, we can still discern a fittingness in the fact that a lifetime of pursuing empty pleasures reduces one to destitution. And this, once more, reveals God's undying love: that God allows the sinner to grasp the connection between sin and madness, and thereby grants her the opportunity to turn away from her sin.

We must pause for a moment to appreciate the apparent *contingency* of this arrangement, in which the form of madness is a mirror of, or more generally *displays the truth of*, the corresponding vice that birthed it. It didn't have to be this way. God *could* have designed reality in such a way that sin always meets with a punishment that in no way, shape, or form resembles it. When God struck down Nadab and Abihu for bringing unauthorized incense, there was no sense in which the punishment—being struck dead immediately—was a portrait of their specific act of disobedience. Almost all forms of disobedience were met with death. God's punishment, rather, was simply a way of expressing that their disobedience would not be tolerated. Disobedient people would simply be killed or exiled; they had no place in the community of Israel. So, though Nadab and Abihu's punishment reflected God's justice, there was no sense in which the *form* of the punishment represented God's mercy. God did not provide them with an opportunity to recognize the nature of their sin and to redeem themselves. This feature of madness that Burton discovered, this mirroring function, is one that should inspire a profound sense of gratitude in our hearts. Through it, God deals mercifully with his wayward children.

III.

In some ways, then, it is more illuminating to read Burton alongside the morality plays of his day—for example, the so-called Macro Plays of the early fifteenth century, or even didactic poetry such as *The Divine Comedy*—rather than alongside Renaissance medical treatises. By presenting us with a catalog of the various forms, or ways, of man's erring, and their consequences, the chapters of the book stand as so many warning signs to stay the course of righteousness.

Consider the Macro Plays, such as *The Castle of Perseverance*, probably penned in the early fifteenth century. The characters in this play include Gluttony, Lechery, Sloth, and Mankind. In the play, Gluttony, Lechery, and Sloth conspire to trick Mankind, and they succeed in their vile task:

> In glotony, gracyous now am I growe; perfore he syttyth semly here be my syde; In lechery and lykynge, lent am I lowe, & Slawth, my swete sone, is bent to abyde.[24]

Soon enough, Mankind is destitute, his possessions having been given away to someone he does not even know. At last, Mankind dies and the bad angel carries him to hell. In the next scene, we are transported to the throne room of God, where we witness a lively debate between the virtues—Truth, Justice, Mercy, Peace—each pleading its case before God. Truth and Justice demand that Mankind remain in hell to pay the just penalty for his sin, while Mercy and Peace plead for his redemption. The play closes as God sides with Mercy and Peace and Mankind is lifted to heaven.

Alternatively, consider *The Divine Comedy*. Here, Dante anticipates the various forms of punishment in store for various sorts of sins. In some cases, the forms of punishment that Dante envisions exhibit this apt *fittingness* to the corresponding sin. For example, those who have indulged repeatedly in the sin of lust find themselves, in hell, whipped about by fierce winds; this is a suitable punishment for those who, back on earth, allowed their base impulses to whip them about.

Cheyne, writing a century later, still accepts, like Burton, this dual teleological picture. His purpose in *The English Malady*, is not, in the first place—and here we must take a stance on the history of science proper—to expose melancholy as a disease spawned by nerve damage. More fundamentally, his goal is to expose melancholy as a disease spawned by rebellion against the *place-based providence of God*. To understand the mechanism of melancholy one must grasp, in the physiology of the nerves, *that* and *how* God has implanted within us a series of progressively catastrophic warning signs, designed to alert beings like us, *finite fallible lapsed intelligences*, to the full extent of our waywardness.

4

An Infinitely Wise Contrivance

The early eighteenth-century physician George Cheyne was one of the last theorists to openly and unabashedly endorse this dual teleological conception of madness. In particular, he sought to reveal melancholy—and its lesser forms of hypochondria and hysteria—as instrument of both punishment and redemption (Section I). But what, precisely, is the sin *for which* melancholy is a punishment? It is the sin of intemperance, that is, overindulgence in food and drink (Section II). But this evokes a mystery. Of all the morally heinous acts that people commit, why is intemperance *especially offensive* to God? This question cannot be answered unless we read Cheyne's most famous work, *The English Malady* of 1733, in the broader context of his works of natural theology. In short, God has given to the peoples of each region of the globe precisely the kinds of food and medicine that *those* people need to flourish, *given* the demands of their specific habitats. Intemperance is a way of defying God's place-based providence (Section III). What is extraordinary about Cheyne's vision is not merely that he embraces the dual teleology of madness, but that he discerns, in the human frame, a number of exquisitely designed mechanisms—"infinitely wise contrivances"—that function together as an early warning system when we begin to rebel against God's providence (Section IV). Ultimately, for Cheyne, the purpose of madness mirrors the *purpose of embodiment itself* (Section V).

I.

In his *Philosophical Dictionary*, Voltaire tells us that England is the land of suicide:

> We are not informed, nor is it likely that in the time of Caesar and the emperors the inhabitants of Great Britain killed themselves as deliberately as they do now, when they have the vapors which they denominate the spleen. On the other hand, the Romans, who never had the spleen, did not

Madness. Justin Garson, Oxford University Press. © Oxford University Press 2022.
DOI: 10.1093/oso/9780197613832.003.0005

hesitate to put themselves to death. They reasoned, they were philosophers, and the people of the island of Britain were not so. Now, English citizens are philosophers and Roman citizens are nothing. The Englishman quits this life proudly and disdainfully when the whim takes him, but the Roman must have an *indulgentia in articulo mortis*; he can neither live nor die.[1]

Suicide, and the melancholy that birthed it, was popularly known at the time as "the English malady." This epidemic of suicide, or at least the perception thereof, led George Cheyne to publish his masterwork, *The English Malady: Or, a Treatise of Nervous Diseases of All Kinds, as Spleen, Vapours, Lowness of Spirits, Hypochondriacal, and Hysterical Distempers, &c.*, of 1733. In the book's preface, Cheyne informs us that he would not have rushed the book to publication—in fact, he had originally planned to publish it posthumously—but at the urging of friends, he thought that the book might constitute an intervention into the spate of self-murder, and thereby, the book might help "put a stop to so universal a Lunacy and Madness."[2] (A note on terminology: in the eighteenth century, *spleen*—or alternatively, *vapors*—and *melancholy* are sometimes regarded as distinct ailments, and sometimes as different degrees of a single, mutable ailment.)

The English Malady must be read, first and foremost, as a detective story: What is wrong with us? What is wrong with our culture? Whence this madness and lunacy? And how to bring it to an end? Historians sometimes claim that the specificity and originality of Cheyne's thought lie in his viewing melancholy, and its lesser forms, hysteria and hypochondria, as a disorder of the *nerves*. This would situate Cheyne within a wider trend in the history of medicine away from the Greek system of humors to the newer system of nerves promoted by luminaries such as Willis and Cullen. And while this is correct if our only goal is to capture Cheyne's "place" in the history of psychiatry, it is a mistake to say that Cheyne thought of melancholy and its lesser forms primarily as a "disease of the nerves." For while it was nervous, it is not a disease, or not merely a disease. Melancholy is both divine punishment and gift. It is, in fact, an *antidote* to a disease rampant among the English: their intemperance and immoderation in food and drink, and ultimately their rebellion against, and rejection of, God's providence. Melancholy is a crossroads. It is the fulcrum that divides a life into two stages or chapters: the first being rebellion against God, which, for the English, manifests itself primarily through the vice of intemperance, and the second being the restoration, or return, to God, which manifests itself primarily through the virtue of temperance and, more fundamentally, an abiding sense of gratitude for God's generous provision. For these reasons, we must place Cheyne firmly within this madness-as-strategy tradition, if it can, in fact, be called a "tradition."

In melancholy, then, we see Burton's dual teleological structure, this improbable fusion of justice and mercy of which we will say more shortly. But there are crucial differences between Burton and Cheyne. One difference is that, for *Burton*, the forms of madness are universal. When we trace how unchecked anger leads to mania, or grief to melancholy, or covetousness to paranoia, we grasp a universal law of human nature, grounded ultimately in a universal psychophysiological mechanism: *disposition becomes habit*. And while it is true that Burton's England provided unique opportunities for the kinds of moral turpitude that lead to madness, such as overindulgence in gaming or hunting, this fact does not undermine his ambition, in the *Anatomy*, to describe universal human psychology.

The title of Cheyne's book, in contrast, announces a radically different orientation. His is an ontology of the particular, the contingent, the regional and local. "The English malady," for Cheyne, is, as the name implies, a distinctively English phenomenon; it is "a Class and Set of Distempers, with atrocious and frightful Symptoms, scarce known to our Ancestors, and never rising to such fatal Heights, nor afflicting such Numbers in any other known Nation."[3] Cheyne's contemporary Sir Richard Blackmore called it the "English spleen," and noted that it is "comparatively but seldom found among the Inhabitants of other Countries."[4]

For Burton, the very phrase "English malady" would have been an oxymoron, if not a contradiction in terms. Disease is universal, even if the opportunities to acquire that disease vary from region to region. For Cheyne, disease is particularized, stratified and partitioned by culture, character, climate, geography, race, class. One cannot speak of melancholy without speaking of the particularities of England in the early eighteenth century, its being an island nation buffeted by strong cold winds, its moist air, the London fog, the high population density in its cities, the widening gap between social classes, its geopolitical ascendancy. Some theorists of madness insist that if a type of madness is specific to a place, it is to that extent unreal, a mere social construct.[5] For Cheyne, melancholy is real precisely *on account of* its particularity, its boundedness; it is real, not in the manner of the philosopher's natural kind—gold, water, species—but in the matter of an individual: an '83 Plymouth parked in the driveway, Barack Obama, the World's Fair of 1933.

We will resume our detective story. Why is this an *English* malady? Why the spate of self-murder? Why this disease that makes up almost one-third of the complaints of the people in England? Why *us*? Yes, it is true that the malady is related somehow to all of the factors noted above: climate, population density, class, etc., etc. Yet behind these merely empirical factors, these "biopsychosocial" factors, is a constellation of moral factors. Though melancholy is somehow exacerbated or fueled by these empirical parameters, its origin is moral and theological. Put differently, a person of pristine moral character, *if ever there were one*, would not be susceptible to melancholy. The empirical factors, these "biopsychosocial influences," to be effective, must interface with, mesh with, a kind of inherent depravity. In a morally pure person, these empirical factors would find nothing to latch onto.

II.

At the very outset of his book, in the dedication to Lord Bateman, Cheyne gives us a clue to the moral structure of his thought, and one might even say the key to his thought as a whole. Cheyne begins by tracing a series of curious relationships—connections, resonances—between food, writing, and moral character:

> The chief Design of these Sheets is to recommend to my Fellow Creatures that plain *Diet* which is most agreeable to the Purity and Simplicity of uncorrupted *Nature*, and unconquer'd *Reason*. Ill would it suit, *my Lord*, with such a Design to introduce it with a Dedication cook'd up to the Height of a *French* or *Italian* Taste. Addresses of this Kind are generally a Sort of *Ragous* and *Olios*, compounded of Ingredients as pernicious to the Mind as such unnatural Meats are to the Body. Servile Flattery, fulsome Compliments, and *bombast Panegyrick* make up the *nauseous* Composition. But I know that your *Lordship's* Taste is too delicate, and your Judgment too chaste to be able to bear such *Cookery*. Your taking these sheets into your Patronage will probably be a Post not to be maintained without some Difficulty. *Prejudice, Interest,* and *Appetite* are powerful Antagonists, which nothing but good Sense, solid Virtue, and true *Christian* Courage are capable of opposing.[6]

The dedication proceeds through a series of oppositions, and these oppositions align themselves neatly under a moral polarity. The English taste for simplicity and substance is contrasted with the French and Italian tastes, laden with their sauces and spices. *We*, the English, enjoy plain and earnest writing, in contrast to writing that proceeds through flattery and panegyric, and which makes up a "nauseous composition." By calling it a *nauseous composition*, he is instructing us to react to it in the exact same way that we react to toxic food: it should be vigorously rejected, forcefully expelled. (Much later in the book, Cheyne praises the writing style of a Dr. Cranstoun, whose writing is admittedly a bit clumsy, a bit unpolished, but nonetheless "infinitely preferable to all *Varnish*, and shews him equally an *excellent Physician*, and a Man of *Probity*.")[7] Finally, he sets those who, like Lord Bateman, "fight under the Banner of Truth," who exhibit "Temperance, Sobriety, and Virtue," against those who are moved only by "Prejudice, Error, or Malice."

Two questions immediately arise. First, what is the connection between writing, food, and moral character? More to the point, what is this connection *such that* Cheyne can describe the connection as if it were blindingly obvious, as if it required no further elaboration, justification, or explanation, as if any reasonable and well-intentioned reader would be expected to nod in silent approval? Second, how does this complex series of identifications and oppositions reveal something essential about the English malady, not only of its cause and its treatment, but its secret nature?

A conjecture that appears reasonable at first glance, but is ultimately quite superficial, is that the series of oppositions all circle around the underlying value of *truthfulness*. Drenching food with sauce is a way to conceal the taste of the food, and to that extent, it prevents one from making a truthful, veridical, judgment of its taste; it distorts one's ability to discern the gustatory properties of the food. By the same token, flattery and praise are abdications of truth; they are meant not to communicate information accurately but to puff up one's addressee. From there, the connection to moral character is obvious. Flattery is a form of dissimulation that is usually carried out in the interest of a ruling passion, not in the interest of truth. This leads to the conjecture that truthfulness, that is, the opposition between truth and falsity, is an organizing principle for Cheyne's text.

Tempting as this line of thought is, we must firmly reject it. Not only does it *not* help us make progress on answering our question—melancholy, after all, is not a malady of dissimulation—but it is contradicted when we situate *The English Malady* within Cheyne's corpus as a whole, the key principle of which is *providence and its rejection*, not *truth and its dissimulation*. And what we find is that Cheyne's moral invective is not, in fact, targeted against the French and Italians, but against the English appropriation of a form of cuisine that God never intended them to enjoy.

III.

Chronologically, *The English Malady* falls between two monumental volumes, the youthful *Philosophical Principles of Religion* of 1705 (the second edition of which was published in 1715) and the *Essay on Regimen* of 1740. The title of the latter should not mislead: it is not an essay on regimen. Both are extraordinary works of natural theology: their purpose is to demonstrate how God's existence, God's power, and God's kindness can be known by thoughtful observation of the natural world, from the universal scale of stars and galaxies to the microscopic scale of cells and organelles, and all of the realms in between—ecology, geology, physiology. *Philosophical Principles of Religion* is an extremely learned tome; Cheyne was friends with scientific and philosophical luminaries such as Newton, Hooke, and Hume. In any direction we look, we find not only natural demonstrations of God's existence and his power, but also his concern for the welfare of humanity.

Consider the seasons. Why the progression of the seasons, spring, summer, autumn, winter? If we had but one season—say, summer—the ground would be exhausted quickly from bringing forth vegetation, and "in a few years the *Earth* wou'd have been reduc'd into a *Wilderness* of needless Herbs."[8] In a perpetual summer, "there wou'd not fall sufficient Quantities of Rain, to moisten and soften the Mould, to that degree that it is necessary for constant *Vegetation*."[9] Animals could survive neither a perpetual summer, nor a perpetual winter, for their vigor, and the well-functioning of their circulation and nerves, depends on the rhythmic alternation of cold and heat. But to ensure that alternation, God needed to create the axial tilt of the Earth, which is precisely what it needs to be—an angle of 23.5 degrees relative to the plane of the ecliptic—for us to be able to enjoy four seasons. So, the angle of the equator relative to the plane of the ecliptic has a meaning for the welfare of man. The same is true of the Earth's diurnal rotation: people need rest each day so their energy can be restored for labor and recreation. Hence the motion of the Earth, the alternation of day and night.

Now, the tilt of the equator relative to the ecliptic means that not *all* parts of the earth will experience seasons as the English enjoy them; in particular, the equatorial regions will not experience the pattern and tempo of seasons as the English do. The skeptic or atheist might attempt to use this fact to debunk Cheyne's natural theology: "Now perhaps against all these it may be objected, that there are rational Creatures, who inhabit this *Globe* of ours, who are perpetually in both these Extremes, and yet are found to be not at all dispos'd, as I pretend they wou'd be."[10] Still, God has compensated for the equatorial lack of seasons with admirable precision by giving those regions a superabundance of rain during specific parts of the year to ensure that the land remains vibrant and productive. Hence, a fact that the skeptic or the atheist would seek to deploy as a rebuttal or a refutation of God's providence, of his wisdom and power, turns out, on closer inspection, to yield further *confirmation* of it. The beauty of design consists in the interweaving of all things for the good of humanity: big to small, cosmic, global, regional, local. The attitude of a properly trained scientific mind is not sneering skepticism but quiet reverence.

What the progression of the seasons reveals is that *geography, location, place* is an important part of God's moral vision for humanity. God did not make all places the same. England is a small island buffeted by strong winds; France and Italy are temperate; the Near East has vast deserts. It might be a profound mystery why there are *regions*, why places differ, but what is *not* a mystery is why God gave different resources—the appropriate fauna and flora—to those different places. God put camels in the desert so that they could transport people for long distances without needing water. Moreover, God has provided different regions with different foods and different kinds of medicinal plants; it stands to reason that, given his wisdom and munificence, God has provided us with everything we need to thrive in the region in which he has placed us:

> It is highly probable, that the infinitely wise Author of our Nature has provided proper *Remedies* and *Reliefs* in every *Climate*, for all the Distempers and Diseases incident to their respective Inhabitants, if in his Providence he has necessarily placed them there: And certainly the *Food* and *Physick* proper to the middling sort of each *Country* and *Climate*, is the best of any possible for the Support of the Creatures he has unavoidably placed there, provided only that they follow the Simplicity of Nature, the Dictates of Reason and Experience, and do not lust after *foreign Delicacies*: as we see by the Health and Chearfulness of the *middling Sort* of almost all Nations.[11]

This place-based ontology, this bio-theo-regionalistic outlook, helps us understand and clarify the nature of *The English Malady*'s opening polemic against the French and Italian taste. When Cheyne denounces food that is "cook'd up to the Height of a *French* or *Italian* taste," this is not a vulgar ethnocentrism, but a condemnation of the revolting manner in which the English emulate the French and Italian diets. God never intended those foods for the English constitution:

> *The* Diet *and* Manner of Living *of the middling Rank, who are but moderate and temperate in Foods of the common and natural Product of the Country,* to wit *in* animal Foods *plainly dress'd, and* Liquors *purify'd by* Fermentation *only, without the* Tortures of the Fire, *or without being turned into* Spirits, *is that intended by the Author of Nature for this* Climate *and* Country.[12]

We live in a cold, overcast climate. This tends to slow our circulation and make it sluggish. Salty, fatty foods introduce compounds into the blood that tend to obstruct and retard the free flow of the circulation. This is not the case in Italy, where the heat of the sun quickens the circulation: "In the *Southern* Climates, as there is scarce any, at least few, *Nervous* Distempers of the lingering and *chronical* Kind . . . The Warmth and Action of the *Sun*, keep[s] the Blood and Juices sufficiently fluid, the *Circulation* free."[13] Hence, despite the fact that Cheyne speaks often of the nervous system, blood circulation, and so on, *The English Malady* is not, as historian Roy Porter has it, a prelude to "the neurological school of psychiatry."[14] And though Cheyne writes on social class, and though he condemns the rich for importing salty, fatty, foods, and thereby making themselves susceptible to melancholy, this book cannot be read as a salvo in the class struggle. Cheyne tells us only as much about the nervous system as a thoughtful and educated person would need to know in order to be convinced that the English lifestyle is an abomination to God.[15]

IV.

We are, at long last, in a position to understand the series of connections, and the peculiar relationships, that Cheyne articulates in the dedication of his book among food, writing, and moral character.

Now that we have situated *The English Malady* in relation to Cheyne's work as a whole, and in particular, in the valley between the momentous

Philosophical Principles of Religion, on the one hand, and *Essay on Regimen*, on the other, we are in a position to understand melancholy: its cause, its nature, its treatment. Melancholy is caused by a rejection of God's providence, through a restless dissatisfaction with resources that he so generously bestowed upon us for the place which he gave us to inhabit. Its consequence is a reformation. This is not only a reformation of what we eat and how we move our bodies—that kind of reformation is only going to be temporary unless it is fused with an appropriately transformed character. Hence in its innermost nature, melancholy fully and completely exemplifies the dual teleology that we have already witnessed in Burton: it represents that mysterious synthesis of God's justice and God's mercy. *Just*, in that it is a punishment that we deserve; *merciful*, in that, through melancholy, we are given the opportunity to reflect on our lifestyle and to change our character.

Melancholy is a deserved punishment for sin. But we must be extremely cautious in the way that we understand the relationship between God's will and melancholy. God is not the author of melancholy. God does not bestow melancholy on a person in the same way he bestowed madness on Nebuchadnezzar. Rather, given the design of our bodies, melancholy is a natural result of our intemperance and immoderation. *We bestow melancholy on ourselves.* Think of the sword of Ajax that Burton alludes to: first and foremost, melancholy originates from the *misuse* of our faculties. True, God foresaw that the English lifestyle and diet would induce melancholy, and he even designed the causal structure of reality so that it would have just this effect. But God does not, as it were, directly inflict melancholy on any particular individual:

> I can never be induc'd to believe that the omnipotent and infinitely good Author of it, could, out of Choice and Election . . . have brought some into such a State of Misery, Pain, and Torture, as the most cruel and barbarous Tyrant can scarce be suppos'd wantonly to inflict. . . . No! none but Devils could have such Malice; none but Men themselves, or what is next themselves, I mean their Parents, who were the Instruments or Channels of their Bodies and Constitutions, could have Power or Means to produce such cruel Effects.[16]

Rather, because of the very general laws of cause and effect, certain kinds of actions are bound to have negative consequences, but these laws themselves, in their regularity and uniformity, are good. Gravity is a law of nature that is good, but it is also the means by which one man drops another out of an airplane into the sea. The law is good, but the human will is evil, and people use law to further their evil deeds:

> In itself this Law and Establishment of Nature has infinite Beauty, Wisdom, and Goodness: *viz.* by this progressive and continual Succession from one Root, that the Healthy and Virtuous should thereby be growing continually healthier and happier, and the Bad continually becoming more miserable and unhealthy, till their Punishment forced them upon Virtue and Temperance; for Virtue and Happiness are literally and really Cause and Effect.[17]

This passage expresses the dual teleology that we have been investigating, with this crucial difference: while Cheyne recognizes that melancholy is from God, God *never* violates the laws of nature; God harnesses those laws to accomplish his ends. To the extent that it is a punishment, we have to understand punishment *as* natural cause, that is, as outcome of the collaboration of God's good laws of cause and effect and our evil will: hence the first wing of his dual teleology. This is in contrast to Burton, who, though he recognized the manner in which God harnesses the laws of nature to bring about just punishment, still admitted the occasional role of demons, witches, and spirits in the etiology of disease.

Melancholy also embodies God's mercy. For God desires that we, disgusted with our own condition, would mend our ways, change our characters, and be grateful for his place-based providence. Unlike the punitive aspect of melancholy, which proceeds from the laws of nature combined with man's evil will, God plays a more *active* role in the process of healing, in the process of recovery, redemption. To fix our ideas, however, we must understand the divide between law and mechanism.

At the turn of this century, philosophy of science witnessed a significant reorientation in its thought about the ultimate goal of science. We can put this, somewhat crudely, in terms of a transition from law to mechanism.[18] According to this law-based view, the mission of science is to discover the *laws of nature*. A law is, first and foremost, something that can be stated as a universal, exceptionless, generalization: the law of gravitation, for example, states a relationship between a handful of parameters that is perfectly general. The law of refraction states a general principle about the way light interacts with surfaces. Philosophers still debate vigorously whether there are *laws*, in the proper sense, of ecology, of sociology, of economics. In addition to laws, or as an alternative to laws, we can look at science as having for its vital mission the discovery of *mechanisms*. We speak of a mechanism of digestion, a mechanism of blood circulation, a mechanism of cross-fertilization. In this conception, the goal of science is to discover mechanisms and find out how they work. Mechanisms, unlike laws, are rife in the living world, for a reason we'll come to understand shortly.

Mechanisms and laws reflect fundamentally different construals of the nature and purpose of science. Mechanisms, unlike laws, are *particular*: the mechanism of digestion in a human is unlike that of a cow, which is unlike that of an insect, though they share certain similarities. The way to discover a law is to demonstrate a regular correlation between two or more variables: pressure, temperature, volume. The way to discover a mechanism is to take it apart, to fiddle with the pieces, to break it. The fact that there is a difference between laws and mechanisms is salient even to the philosophically untutored ear: while it is natural to speak of a "mechanism for digestion," it is unnatural to speak of a "law of digestion."

There is another feature of mechanisms that sets them off decisively from laws: a mechanism has an *end*, a *purpose*, a *reason for being*; a mechanism is partly defined by, constituted by, the end that it has the purpose of producing; the outcome of a mechanism, its proprietary phenomenon—blood circulation, dopamine homeostasis, digestion—is at the same time, as a rule, its purpose and its reason for being. The teleological character of mechanism is embedded deeply in biological thought: mechanisms are usually mechanisms for ends that we consider to be good, useful, beneficial, adaptive, in some manner. *Diseases, as a rule, do not have their own mechanisms*; they result from the *breakdown* of a mechanism for a good, functional, healthy state.[19] Drug researchers often describe drug addiction not as "having" a mechanism, but as resulting from the breakdown of a mechanism for reinforcement learning. Alzheimer's disease has no mechanism of its own; it results from a breakdown in the mechanism of memory. From this perspective, it is puzzling that mechanism has traditionally been thought to be somehow opposed to teleology: mechanism realizes teleology.[20] It is unthinkable without teleology.

For these reasons, it would be a mistake to treat the theological framework of Cheyne's text as a dispensable shell, a kind of eccentric side project, much like the way we treat Isaac Newton's preoccupation with astrology when we discuss his scientific achievements.[21] Though we might be correct enough to do so in the case of Newton, it is entirely out of place here, for the reason that we cannot understand Cheyne's physiology without understanding the theology that motivates and informs it. The way that Cheyne identifies the mechanisms of the human frame, their composition, their behavior, cannot be divorced from his prior grasp of God's specific purposes. My simply being able to recognize that within the human body, in addition to a mechanism for digestion, for circulation, and so on, there is a mechanism that has the function of causing us to refuse food and drink when we are sick—my ability to identify that *as* a distinct mechanism among others, with its characteristic composition and behavior—depends on my prior grasp of God's moral vision for humanity. This, in short, amounts to restating a principle that has been formulated repeatedly throughout the last century, that of the ineliminability of teleology from biology. As Mayr argued, the most sensible starting point for physiology is, *what is it for?*[22] In Cheyne's time, this *what is it for* was understood in terms of divine intention; today, we understand it in terms of Darwinian natural selection ("why did natural selection put it there?").

This rough and preliminary overview of laws and mechanisms gives us a foundation for understanding the way in which God is fundamentally and directly involved in healing: he has designed and placed *mechanisms* in the human body whose sole purpose is to help us to resist a lifestyle of immoderation and intemperance, and to initiate the moral transformation that is a precursor to healing. In his wisdom, God implanted a mechanism, an "infinitely wise contrivance," such that when we begin to experience digestive problems owing to our licentious lifestyle, we experience a natural inclination to abstain from food and drink:

> There is also another *infinitely wise* Contrivance in Nature, that *Loathing* and *Inappetency*, or at least a Difficulty in Digestion, always attends, in some Degree or other, all Disorders whatsoever. Were every one that is a little ill, capable of the same *Riot* and *Excess* during their Distemper that they were when in perfect Health, when they laid in the Materials of their Disorders, they would infallibly and quickly ruin themselves, and perish without Resource.[23]

In doing so, God reveals to us, even the uneducated, that the best road to health is temperance:

There is no surer or more general *Maxim* in *Physick*, than that Diseases are cured by the contrary or opposite Methods to that which produc'd them . . . there needs no great Depth of Penetration to find out that *Temperance* and *Abstinence* is necessary towards their *Cure*.[24]

He clarifies the point in *Essay on Regimen*:

It is enough, for a *wise Man*, and a *Christian Philosopher*, to stop, and attend to *Diet* and *Regimen*, when, by the *Order of Providence*, his natural and providential Course of common *Regimen* is *barr'd up*, by some Disorder or Distemper; and then wise Nature will give him timeous Warning, by *Inappetency*, a *Nausea*, Reaching, *Vomiting*, a Flatulence, Fulness or Pain in the Stomach; for all Distempers begin first at the Stomach or Bowels, and then ascend to the *Head*, which is the Language of the *God* of *Nature*, saying to the Person, *Man, take Care*.[25]

As far as God's active involvement in the process of healing, God has placed within us not just a single warning sign (inappetence), but a whole *series* of warning signs. God does not allow the disease to possess us *all at once*, in its fullness. God has ordained, rather, that in the course of a licentious lifestyle, there will appear a series of *progressively severe physical reactions* or crises, each of which constitutes both a warning and an invitation. A person afflicted with the vapors exhibits *three degrees* of progressive deterioration. The first degree is marked by indigestion, chills, flushing, burning, sweating, and lethargy; the second by the same symptoms "in a much higher and more eminent Degree, and some new ones," including a "deep and fixed *Melancholy, wandering and delusory Images* on the Brain, and *Instability* and *Unsettledness* in all the intellectual Operations"; at last, the third and final degree is marked by some "*mortal* and incurable Distemper."[26] Therefore, anyone who tastes this final degree of deterioration possesses, by that fact, an almost unthinkable stubbornness, an inconceivable hardness of heart. It is impossible not to think of the book of Revelation, which describes not only the horrific penalties in store for God's enemies, but *what is even more horrific*, their repeated refusal to repent of their misdeeds, even when God, in his mercy, grants them opportunity after opportunity to do so.[27]

V.

In all of God's machinations, in his mechanisms, his contrivances, his warning signs, he never violates the liberty of his created beings, but he engages with humanity in such a manner as to respect the free exercise of the human will, and more generally, the will of intelligent created beings, whether humans or angels: God, "by an over-ruling *Providence*, and, as it were, by meer *casual Hints*, far beyond the Reach of my *Penetration*, has irresistibly (as I should almost say, if I felt not my own *Liberty*) directed the great Steps of my *Life* and *Health* hitherto."[28] As it turns out, this dual teleological structure of madness, madness as punishment and opportunity for redemption, is a microcosm of the very nature of embodied existence. This is the thesis expounded at some length in his demanding 1740 treatise, *Essay on Regimen*. Humanity, during this phase of its existence, lives in a valley between two events: the Fall and the Judgment. The physical constitution of the world, and of our bodies, is conditioned by the Fall: that first act of defiance against our creator. That act not only erected a spiritual wall of separation between humanity and God, but it altered our physical reality as well. Our very bodies became susceptible to disease and death. Childbirth became painful. We must conclude that the bodies we occupy now are not entirely like the bodies we occupied before the Fall. Prior to the Fall, we had bodies, to be sure, but these bodies were made of a more subtle form of matter; they were more pliant and supple, more responsive to the will. The bodies we inhabit during this phase of human history are composed of a grosser, denser matter, subject to disease and accident; they are no longer entirely obedient to our will. They are, in short, prisons. We have been imprisoned for this particular era of human existence:

> When human Nature had thus *lapsed*, by affecting Independence, and desiring to govern itself by its own natural Spirit solely, and wallowing in the Objects of Sense . . . the human Body did hence necessarily and *mechanically* (as it were) contract a Rust, Grossness, *Stupor* and Inactivity, and became restive and disobedient to the Commands of the natural *Spirit*, gradually degenerating into an *earthly, gross, material* Prison or *Dungeon*.[29]

This prison of embodiment serves two functions, expiation and purification: *to punish and to redeem*. The condition of embodiment itself embraces these two aspects of God's character, his justice and his mercy:

Infinite Wisdom and Power . . . seems to have contrived this wonderful expedient, *viz.* to tye down, sopite and restrain the Acts and Exertion of the
natural Powers, of laps'd, *sentient* and *intelligent* Beings, for a determin'd
Space of Time, by Chains and Fetters made of the *Elements* of this ru-
inous *Globe*, in order to punish and purify them, and so to vindicate his
Sovereignty, to repair the indignity done to his *Purity*, to warn and deter the
other Orders of his standing *Hierarchies*.[30]

It is clear how these gross bodies constitute a *punishment* for disobedience,
and perhaps even a *warning* to other created intelligences, that is, angels. But
how do they constitute an *opportunity for* purification? They do that because
along with our weakness comes our recognition of the extent to which we,
created beings, rely on our creator; that recognition moves us to seek his forgiveness and healing, and in that act, in our full and sincere submission to his
sovereignty, the right relationship is restored and the long journey toward
health begins. The passage continues:

And at the same time, by lessening the Strength and Activity of the natural Powers in the full Vigour, to allow Freedom and Uninterruption from
them, for the Restoration and Advancement of the moral Powers.[31]

Hence madness must be seen as a *punishment inside a punishment*, one that
serves to gently guide us, to nudge us along in our journey of purification.
But this nudging, this guidance, cannot be thought of in terms of compulsion
or coercion. Part of the beauty of God's plan for humanity lies in the deli-
cacy of the relationship between God's plan and human volition. For God,
in his dealings with us, and in the matter of our purification, always acts in
a way that respects the free exercise of our will. God does not compel or co-
erce a person to turn to him or put their faith in him, for coercion would
negate the very purpose, the very possibility, of our purification: developing
moral character requires developing virtues such as faith and trust, virtues
that, by their very nature, cannot be compelled. This is one reason that God
has chosen not to reveal himself in an unmistakable manner to humanity, the
"hiddenness of God": it would amount to a form of mass coercion.

But to say God "hides himself" does not mean that God leaves no trace of
his existence in the natural world. As Paul says, "For since the creation of the
world God's invisible qualities—his eternal power and divine nature—have
been clearly seen, being understood from what has been made, so that people

are without excuse."[32] Human experience furnishes numerous lines of evidence for God's existence and characteristics. Still, God has allowed for a kind of *epistemic gap*: it is possible for a reasonable creature to disbelieve in God; God's existence is not evident in the same way as a leaf or cloud is. But this epistemic gap—the fact that God has chosen not to reveal himself in an unmistakable way to all—is by no means an error; it is a design feature, a fundamental prerequisite of cultivating the virtues of faith and trust:

> Unerring Evidence, irrefragable Demonstration, absolute Certainty, must necessarily interfere with *Humility, Dependence, Resignation, Faith* and *Trust*, and consequently with all Merit, Gratitude, and Love. What Faith? What Resignation? What Merit is there in believing the Propositions of *Euclid*?[33]

In short, the purpose of madness is, at root, the same as the purpose of embodied existence itself. It serves to punish and to purify but in a manner that does not violate the sovereignty of our will. Madness, *in one sense*, is not a disease, but a cure, an antidote; or, better, it is as much a disease as it is the cure for the disease of intemperance. But if madness is the cure of the disease of intemperance, then it is possible to conceive of madness as a form of sanity, a *higher*, truer sanity. Sanity, at least that sort of "sanity" with which we are intimately familiar—the kind of frenetic impulse to acquire, to chase luxury, to consume, to live beyond our means—is madness. Hence, in Cheyne, we encounter a theme that we will come to see repeatedly throughout the nineteenth and twentieth centuries. This is the *grand inversion* that makes madness sanity and sanity madness:

> From all which is it evident, that these monstrous and extreme *Tortures* [of melancholy], are entirely the Growth of our own Madness and Folly, and the Product of our own wretched Inventions, from the Poison and Ordure, with which, for the sake of a little sensual Pleasure, we forcibly and tyrannically cram our poor passive Machins.[34]

This inversion will recur frequently in psychiatry's history, its most articulate modern spokesperson being Laing: if the fruit of sanity is the Vietnam War, G.I. Joe, better dead than Red, the era of mass incarceration, then madness can be seen as the *refusal to engage*, the rejection of those societal values, and hence what we call "madness" is the true sanity, the higher sanity, the sanity that denounces and renounces the societal madness *falsely known as* sanity. In Cheyne, melancholy cannot be conceptualized as a form of dysfunction; it is by design; it is designed to push back against societal madness, and to initiate the long journey toward healing.

PART II

MADNESS AND
THE SOUND MIND

By the middle of the eighteenth century, we encounter a new consciousness of madness, a new way of "studying the mad." Madness is no longer thought of as divine strategy with the two-fold goal of punishment and redemption. We are too far along on the path to Taylor's secular age to continue to entertain such fantasies in a serious way. Madness is *mere* pathology: when someone is mad, it is because *something has gone wrong inside of them*, in their body or in their mind, and the purpose of medicine is to correct this dysfunction, to make the wrong right, if that is still possible. Madness has no truth of its own to reveal; it no longer speaks to us of sin, or punishment, or justice, or mercy, or God's glorious plan for the salvation of the world. It merely points, mutely, to the broken mechanism from which it originated.

True, we see auguries of this new consciousness of madness, this way of studying and treating the mad patient, even in ancient times. The Hippocratic author of *On the Sacred Disease* believed that all forms of madness spring from dysfunctional flows of blood or brain afflictions. But a decisive shift has taken place; there is yet something in the eighteenth century worthy of the appellation "new." For here, we witness what appears to be an unprecedented style of reasoning about the mad: in order to identify and enumerate the diverse forms of madness, in order to individuate, number, and label them, *one must begin by surveying the faculties of the sound mind*—such as reason, imagination, memory, judgment, abstraction, will, perception. That is because *there are as many forms of madness as there are ways that the faculties of the sound mind can break*. One form of madness is an aberration of reason; another, an aberration of abstraction; a third, an aberration of will. This is the fundamental principle of both classification and treatment. It is for this reason that we begin with Kant: although Kant's youthful "Essay on the Maladies of the Head" of 1764 is almost completely unknown, even within philosophy, and though it made no material impact on the history of psychiatry, it captures this new mode of consciousness with unparalleled rigor, simplicity, even beauty.

Still, we encounter a source of tension that continues throughout the nineteenth century. *On the one hand*, we have madness-as-dysfunction, this project of rethinking madness and all of its forms as nothing but dysfunctions in the corresponding faculties of the mind. *On the other hand*, many of the theorists of this era, including Kant, Haslam, Wigan, Heinroth, Pinel, and even Griesinger, *repeatedly uncover design inside of madness*. It is as if, despite their best efforts to stamp it out, teleology persists; it cannot be entirely canceled or negated.

But teleology, of course, has different sources. So, while these thinkers clearly recognize design in madness, they construe the source of this teleology quite differently. The question, then, that we can pose to each of them is: *In what respect does madness reveal design? And what is the source of this design?* Does it come from the purposes of God? Does it come from the goal-directedness of a mysterious vital force buried within nature? Does it come from the purposiveness of the unconscious idea? Does it come from conscious intention?

For Kant, though all of the forms of madness represent dysfunctions of the faculties of the sound mind, he still is able to discover design, or what he calls the deeply implanted *wisdom of nature* (*Weisheit der Natur*).[1] Of the highest form of madness, *Aberwitz*, he tells us, "It is astonishing, however, that the powers of the unhinged mind still arrange themselves into a system, and that nature even strives to bring a principle of unity into unreason, so that *the faculty of thought does not remain idle*."[2] Even in the extremes of the "unhinged mind" we still discern the working out of a secret plan and purpose.

John Haslam, apothecary to Bedlam hospital at the beginning of the nineteenth century, sees the core problem of madness rather differently than Kant: if there is reason inside of madness, then how is madness possible? What, precisely, must be the relationship between madness and reason *such that* the madman can possess some measure of reason, yet still be mad? Haslam reconciles this tension through his principle that *the madman rejects the claims of reason, just as the sociopath rejects the claims of morality, but he still utilizes reason in order to further his project of dissimulation*. In short, the goal of madness is to conceal and perpetuate itself under the shell of reason.

Johann Christian August Heinroth, the first thinker to hold a chair of psychiatry in Europe, also appears to be among the first to develop, seriously and systematically, the idea that some forms of madness represent *coping strategies*. Madness is a contrivance; through madness, the mind enters into a fictional world, to buffer itself against the horrors of the real one. It is a refuge and a castle. Heinroth's idea that madness is a mechanism for coping with trauma becomes a constant refrain in early twentieth-century psychoanalysis, and particularly in the work of Frieda Fromm-Reichmann and Harry Stack Sullivan, both of whom sought to apply psychoanalytical principles to psychosis.

For Arthur Ladbroke Wigan, madness is a result of what he calls the *duality of the mind*. In his view, each of the brain's two hemispheres is, in fact, a distinct brain, equipped with its own distinct mind. Madness results from the interplay between the two minds, *one of which is healthy and the other sick*. The different forms of madness are merely so many different stances, or attitudes, that the healthy brain adopts with respect to the sick brain. In some cases, the healthy brain strives to assert *dominance* over the sick brain; in others the healthy brain is *deceived* by the sick brain; in yet a third sort of case, the healthy brain *acquiesces* to the sick brain. Hence all madness is accompanied by a purposiveness, a *striving*: the madman is either striving to conquer his own madness, or he has renounced the fight altogether and acquiesced to his madness, *either because his madness deceived him or it offered him something that his reason could not give him.*

Philippe Pinel, chief physician at the Bicêtre in the aftermath of the French Revolution, recognizes that many forms of madness result from the breakdown of a corresponding faculty of the mind. However, incredibly enough, he maintains that some forms of madness—what he calls *manie périodique ou intermittente*, periodic attacks of maniacal fury—have a cathartic and healing function. They represent the intrinsic healing tendency of nature itself, the ancient *vis conservatrix et medicatrix naturae*. Consistent with this is Pinel's advocacy of a non-interventionist philosophy of healing, in which nature is allowed to run its course, with the exception of the most recalcitrant cases.

Finally, Wilhelm Griesinger, at the outset of his *Die Pathologie und Therapie der psychischen Krankheiten* of 1845, sets out methodically his view that *mental disorders are nothing but dysfunctions of the nervous system*. Still, even Griesinger cannot help but find teleology and purposiveness inside some forms of madness, particularly in the delusional disorders. As a rule, *delusions, like dreams, are wish fulfillments*. They have the purpose of changing reality from a place of pain and torment to one of pleasure and hope. Why, for example, do erotic delusions, or delusions of grandeur, get "fixed" more easily than paranoid ones? Because the mind more eagerly embraces them.

If Griesinger's textbook marked the beginning of the era known as "German imperial psychiatry," Emil Kraepelin marked its end. Kraepelin managed to develop and extend madness-as-dysfunction more rigorously, more thoroughly, *more unforgivingly*, than any other thinker so far encountered. For Kraepelin, there is simply no room for teleology in madness. That is because, if we can see goal-directedness and purposiveness inside madness, *we are no longer dealing with madness but malingering*. For what would be the source of this teleology, *if not conscious intentionality*? Neither God, nor vital forces inherent in nature, nor unconscious ideas have any role in his incipient science. So, where else would purposiveness come from, but from conscious human design?

In short, the nineteenth century, the era that, *by all rights*, should have marked the permanent victory, the supremacy, the utter hegemony of madness-as-dysfunction, is repeatedly punctured by madness-as-strategy. Almost all of the thinkers who attempt to establish that the forms of madness are nothing but dysfunctions cannot help but recognize design. A bold exception here is Kraepelin, who maintained, until the end of his life, that mental disorders are brain dysfunctions and that *to the extent that we discover design inside of madness, it is not really madness after all*. For better or for worse, Kraepelin's efforts to establish madness-as-dysfunction ended in frustration: in 1899, in the same year he published the groundbreaking sixth edition of his textbook, Freud published *The Interpretation of Dreams*, which situated the study of madness in an unabashedly teleological context. At the beginning of the twentieth century, the battle was lost: all madness was by design.

5

A Temporary Surrogate of Reason

Kant proclaims a fundamentally new way of conceiving of madness itself, and, at the same time, a method for classifying its forms. Each form of madness is a defect in a corresponding faculty of the sound mind. There are as many forms of madness as there are faculties (attention, abstraction, memory, ratiocination, etc.). This is not merely madness-as-dysfunction, but madness-as-dysfunction filtered through faculty psychology (Section I). Kant's youthful "Essay of the Maladies of the Head" of 1764 sets out this picture elegantly: there are three faculties of the mind (experience, judgment, and reason), and hence three possible ways of being mad (Section II). It is somewhat astonishing, then, that thirty years later, in his *Anthropology*, Kant *rediscovers* teleology inside of madness. The forms of madness do not merely represent dysfunctions of the various mental faculties; at least some of them, in addition, express *the wisdom of nature. Aberwitz*, for example, that pinnacle of insanity, still participates in reason; through it, nature ensures that even the mad attempt to systematize reality, "so that the faculty of thought does not remain idle" (Section III). Having understood Kant's system of classification, it becomes easier to appreciate what is so distinctive about Locke: Locke does not think that the forms of madness represent dysfunctions of the various faculties of the mind. Rather, they represent well-functioning faculties that have set themselves to work on faulty "inputs" (Section IV).

I.

By the end of the eighteenth century, we witness a new mode of conceiving madness, of interacting with the mad person, of speaking to and about the mad. Madness begins its remarkable journey from an expression of divine punishment, a revelation or betrayal of individual sin, to a biomedical phenomenon, and ultimately, to an empirical fact among other empirical facts: it is 85 degrees Fahrenheit out; the Pacific Ocean reaches a depth of over 10,000 meters in places; Jayani is a paranoid schizophrenic; Roby has melancholy.

Madness. Justin Garson, Oxford University Press. © Oxford University Press 2022.
DOI: 10.1093/oso/9780197613832.003.0006

In short, psychiatry begins to assume the form of an empirical science. Almost every psychiatric textbook of the nineteenth century starts by bemoaning the fact that, in the science of madness, empty speculation reigns over studious, patient, even tedious observation of the mad. But this complaint is disingenuous: first, psychiatry was scarcely thought of as an empirical science prior to the eighteenth century, and second, the "Great Confinement," of which we have heard so much, does not actually begin until about the middle of the nineteenth century, and so we simply do not have the *material*, the fodder, for the observation required to make the study of madness into an empirical science.[1]

The thinker who announces this new framework, and who best embodies this new movement in its starkest form, in a simple but logically rigorous manner, is Kant. Kant does not play strongly into the history of psychiatry proper. He did not materially impact psychiatry as did, for example, Locke. Contemporary textbooks of psychiatry and its history make no mention of his work. But this is often enough the philosopher's lot: to articulate rigorously, and without much acclaim, a vision that is implicit in one's scientific milieu. This vision of madness is one that is progressively elaborated from within psychiatry during the nineteenth century, and it is reaches its zenith in the Research Domain Criteria (RDoC) project of today. Though superficial, it would not be a stretch to gesture at the direction of Kant's thought by using the term "medicalization." The problem is that we still do not fully understand what "medicalization" is, the change of vision that "medicalization" is supposed to denote. Given this, the term can only interfere with our ability to trace the outlines of this new movement.

In short, Kant's seminal observation is this: each form of madness represents a dysfunction of a corresponding faculty of the mind. Just as the faculties of the mind form a complete and closed system, one that can be grasped a priori, so too do the basic forms of madness constitute a complete and closed system, whose outline can be grasped a priori. This is indeed a new thing. The fact that the forms of madness can be deduced in an a priori manner does not imply, however, that psychiatry is not an empirical science. Empirical science begins with an a priori stratification of its domain. And though this a priori stratification, this "framework" or "paradigm," can be modified or transformed through empirical science, it precedes empirical science; it is a requirement of empirical science, as a point of departure and orientation, as Ludwik Fleck and later, Kuhn, demonstrated.[2]

Certainly, the idea of madness as involving a dysfunction or deviation from a purpose is not new. As early as Hippocrates, we see the forms of madness as involving just so many dysfunctions of the normal flow of air in the veins or the proportionality of the humors. In Burton, we see the forms of madness as so many departures from the virtuous life. But with Kant, we see two fundamentally new ideas fused together: first, the forms of madness as aberrations of the faculties of the mind, and second, the forms of madness as constituting a complete and closed system—just as the faculties of the mind do.

This Kantian framework gives us a new key for unlocking the secret of madness. Every "madman"—Locke's preferred terminology—has his secret. His secret is this: *which faculty inside of him has failed to perform its function?* The healer's job is to observe the madman assiduously until he delivers up his secret. When I have discovered which faculty is disturbed, when I have seen that your madness is an aberration of your faculty of judgment, say, or your faculty of experience, or of reason, or desire, I have wrenched this secret from your seemingly chaotic and inexplicable thoughts, feelings, and behaviors. I have *situated* the madman; that is, I have identified his *place* in a system of forms. At this point, I can surmise that with the advance of treatment and technology I will be able to assist him, or at least understand how best to manage him.

Kant's system, no doubt, leaves many empirical questions open, even after we have grasped the essence of madness and the finite number of forms that it assumes. But these questions are the very questions with which empirical science commences: What initiates madness? What are its underlying biological mechanisms? What is the ordinary course of the various forms of madness? Their prognosis? Their treatment? Their characteristic symptoms? Average age of onset? The questions we now pose of madness are fundamentally Kraepelinian questions. As important and as valuable, even as necessary, as they are, they are incidental to madness; they are mere details.

Within this new system, madness is truly *silenced*, in a very specific and decisive manner. I observe the madman. I record his speech. I take note of his movements, his intentions, his ramblings, his contraptions, his furious scribblings. At long last, I discover his secret. At this point, madness has no more truth of its own to yield. His words and actions reveal no further truth; his ramblings are now just mad ramblings; they are shells. His contraptions are just mad contraptions. As Foucault put it, madness "cease[s] to be a sign of another world," and is restored "to the truth of its emptiness."[3] If the madman is melancholic and raves that he has sinned against the Holy Spirit and awaits his condemnation, it is because *that is the kind of thing the melan-cholic says*. His productions have value only to corroborate or to disconfirm a diagnosis. I can try to console him; I can try to edify him; I can try to speak kind words to him, but ultimately, I listen to his speech not as speech, but as the byproduct of the inner madness that drives it, that pushes the very words from his mouth. It is Strawson's "objective attitude."[4]

Not only have I silenced the madman after I have discovered his secret, but I have silenced him even before he speaks, and that is precisely because I have resolved, in advance, to listen to his speech not *as* speech but merely as a se-ries of clues to his secret. I track his speech first and foremost in its indicative function, not its expressive function. In fact, the very idea that madness has a truth of its own to reveal is part of the dissimulation of madness. It is a kind of illusion that madness generates to sucker the unwary. In Haslam, this quality of madness, this dissimulating tendency, becomes a defining quality of mad-ness, part of its very essence, its telos. To the extent that Haslam can think of madness as goal-driven and purposeful, it is because he thinks its purpose is to dissimulate, to deceive, and its special mode of dissimulation or deception is precisely to make one think that it has a truth of its own to reveal.

Finally, to the extent that psychoanalysis represents a counterpoint to this Kantian tradition, to the extent that psychoanalysis gives voice to madness, that it "listens" to madness, it is not because it takes a "psychogenic" rather than a "biogenic" point of departure. The distinction between biogenic and psychogenic, or even a compound of those (biosocial, psychobiological, biopsychosocial), is of no further help here. Rather, the difference between the Kantian and the psychoanalytic tradition must be understood in terms of the opposition between dysteleology and teleology. In the Kantian tradition, the forms of madness are dysfunctions or aberrations of the normal facul-ties of the mind, and *therefore* they have no voice, they have no truth of their own. In the psychoanalytic tradition, the madman's words, his actions, his contraptions, his ravings, his scribblings, must be attended with the utmost vigilance because they have a purpose; the madman is desperately trying to solve a puzzle, he is trying to accomplish a goal, and the words and actions represent attempts on his part, or on the part of the "unconscious," to accom-plish that goal.

Kant's approach to classification represents a sharp break from what had been taking place around him, that is, from the efforts of the famous nosologists of his day: Linnaeus, de Sauvages, and most importantly, Cullen. Like Kant, their goal was the classification of the forms of madness, but theirs was *merely* an empirical endeavor, beholden only to the norms of observation and collection. Linnaeus' claim to fame, of course, was the classification of plants, but he brought the same empirical spirit to the classification of madness, or *Mentales*, with its three main genera and twenty-four subspecies. As in botany, nosologists like Linnaeus had no conception that the forms of madness could be deduced from a single principle. Certainly, the basic principle of classification—that one must classify plants by the number and type of stamen and pistil, rather than color or shape—is deduced from broadly philosophical principles. But the adventure, the journey, of discovering the number of plants, cataloging their various forms, and codifying them in a convenient table requires collection and observation. A philosopher who announced that there must be exactly twelve kinds of plants in the world, with such-and-such specific features, would be laughable; she would clearly be transgressing the bounds of her discipline. The systems of Linnaeus, Cullen, and de Sauvages, are fundamentally empirical and open-ended; Kant's is a priori and closed.

Though Kant's system was not widely read and adopted, this Kantian impulse is elaborated through the nineteenth century, as we will see. Haslam, Cox, Spurzheim, Rush—all begin their textbooks in the exact same manner. If you want to "do psychiatry," here is what you must do: you must enumerate the faculties of the healthy mind, and then you must catalog the different forms of madness in terms of the various ways that those faculties can be disrupted or perverted.[5]

II.

Kant begins his taxonomy of madness, in his 1764 "Essay on the Maladies of the Head," modestly enough. There are three faculties of the human mind: experience, judgment, and reason. Consequently, there are three, *and exactly three*, basic forms of madness, and each form corresponds to a breakdown in one of the three faculties: "The frailties of the disturbed head can be brought under as many different main genera [*Hauptgattungen*] as there are mental capacities that are afflicted by it."[6]

The first of these is the disorder of the faculty of experience, *Verrückung*, which Holly Wilson translates as *derangement*. It is simply hallucination, and it takes place when fantasy images are mistaken for "experiences of actual things." It most likely comes about when, due to nerve damage, the brain is stimulated *in just the manner it would have been* were the perception veridical:

> Now let us suppose that certain chimeras, no matter from which cause, had damaged, as it were, one or other organ of the brain such that the impression on that organ had become just as deep and at the same time just as correct as a sensation could make it, then, given good sound reason, this phantom would nevertheless have to be taken for an actual experience even in being awake.[7]

That Kant speculates openly on the biological basis of *Verrückung* does not convert Kant's system into a biogenic one. While the possibility of a biological explanation of *Verrückung* is certainly on the horizon of his text, it is incidental to his system. The biological malfunction *causes* the failure of the faculty of the mind, which alone constitutes madness. Even if the explanation were wrong, even if *Verrückung* were generated in some entirely different manner—for example, even if it had entirely "psychological" causes—it would still be *Verrückung*; it would still be a form of madness, because the essence of *Verrückung* is the inability of the faculty of experience to perform its function.

Verrückung has two lesser forms, hypochondria (*Hypochondrie*) and melancholy (*die Melancholicus*, a person with melancholy). Hypochondria takes place when the illusions of experience relate to one's own body: "The chimeras which this malady hatches do not properly deceive the outer senses but only provide the hypochondriac with an illusory sensation of his own state."[8] Melancholy takes place when these illusions relate to the ordinary trials of life; these trials are exaggerated in an illusory way: "In this regard, the *melancholic* is a fantast with respect to life's ills."[9] Nonetheless, both hypochondria and melancholy are firmly on the sanity side of the gap between sanity and madness; they are not yet madness proper.

Although *Verrückung* is a form of madness, it does not afflict reason itself. One in the grip of *Verrückung* can reason perfectly well, sometimes strikingly so ("given good sound reason, this phantom would nevertheless have to be taken for an actual experience"). The problem is that the experimental premises from which the person reasons are false. One's inferential capacity is unaffected: "Provided one accepts the reversed sensation as true, the judgments themselves can be quite correct, even extraordinarily reasonable."[10] This, of course, echoes Locke's celebrated dictum that the mad reason correctly from false premises.

Next, we get to *Wahnsinn* and *Wahnwitz*, which Wilson translates as *dementia* and *insanity* respectively. *Wahnsinn* affects the faculty of judgment, *Wahnwitz* reason itself. A person afflicted by *Wahnsinn* experiences reality correctly, in terms of the spatial and temporal organization of the world. Such a person, however, imputes false motives to those in their surroundings. The madman is the object of persecution, or secret adoration:

> The *demented person* sees or remembers objects as correctly as every healthy person, only he ordinarily explains the behavior of other human beings through an absurd delusion referring to himself and believes that he is able to read out of it who knows what suspicious intentions, which they never have in mind. Hearing him, one would believe that the whole town is occupied with him.[11]

Wahnwitz, in contrast, is tantamount to full-blown madness. Here,

> all kinds of presumed excessively subtle insights swarm through the burned-out brain: the contrived length of the ocean, the interpretation of prophecies, or who knows what hotchpotch of imprudent brain teasing. If the unfortunate person at the same time overlooks the judgments of experience, then he is called *crazy [aberwitzig]*.[12]

Despite its modesty, Kant's system, as set out in his "Essay," presents madness-as-dysfunction in a simple, elegant, even beautiful way. *Anthropology from a Pragmatic Point of View*, first published in 1798, elaborates, but also in some ways *reverses*, this dysfunction-centered orientation. For here, Kant rediscovers, as if against his own wishes, teleology inside madness.

III.

It would be tempting to read Kant's remarks on madness in the *Anthropology*, over thirty years later, as *merely* an amplification and elaboration of his thinking in the "Essay." Here, he has doubled the forms of madness from three to six, but this is not because he has broken from his understanding of the forms of madness as deviations or disruptions of the basic faculties of the mind; it is just that he has discovered new faculties. These now include the faculties of imagination, of feeling, and of desire.[13] Additionally, Kant has relabeled some of the forms of madness—most likely, as Foucault surmised, to make them cohere better with the nosologists of his day.[14] Now, instead of *Verrückung*, *Wahnsinn*, and *Wahnwitz*, we have *Unsinnigkeit*, *Wahnwitz*, and *Aberwitz*, for the disorders of experience, judgment, and reason, respectively. *Wahnsinn* is the disorder of the imagination. Passion (*Leidenschaft*) and affect (*Affect*) are disorders of the faculties of desire and feeling, respectively.

Still, the continuity between the older system and the newer one potentially obscures an extraordinary reversal. With the *Anthropology*, Kant imports *teleology into madness*, even in a small measure. This is a puzzle. The breakthrough of his early "Essay" consisted in its thoroughgoing renunciation of teleology, and with it, a silencing of the voice of madness. Madness had no truth of its own; it did not reveal reality in a special way; its only "truth" was to point to the pathology that generates it. Now, *affect* and *passion* reveal, in their own manner, providential design, perhaps *in addition to* their being aberrations of mental faculties. They perform a productive function in the life cycle of the individual. Finally, *Aberwitz* itself, that pinnacle of madness, the signature of a "burned-out brain," is not merely a deviation from reason; it *participates* in reason, in reason's systematizing tendency, and this is by nature's design, "so that the faculty of thought does not remain idle."

Begin with affect. Affect is defined as "the feeling of a pleasure or displeasure in the subject's present state that does not let him rise to *reflection*."[15] Affect is not the same as feeling proper. There are healthy and pathological variations of feeling. Affect is a pathological variant. Its pathological nature consists in the fact that in it, the individual is unable to situate her misfortune in the broader context of her experience as a whole. It is a *cognitive* failure; it is a failure of grasping proportionality. The trivial become the momentous. A man is wealthy beyond compare, and his servant accidentally breaks an expensive goblet: the rich man experiences it "as if his entire happiness were lost."[16] This is why affect is not simply feeling; it is feeling at a pitch that obliterates judgment.

But whence the *teleology* of affect? Sometimes affect enables a person to carry out her moral duty even when her reason has not matured to the point of being able to discern that duty for itself, or when reason lacks the strength and maturity to stir up sufficient motivational power to compel action. Affect can act, then, as a surrogate for reason's ability to determine duty before the full maturation of reason's powers:

The principle of *apathy*—namely that the wise man must never be in a state of affect, nor even in that of compassion with the misfortune of his best friend, is an entirely correct and sublime moral principle of the Stoic school; for affect makes us (more or less) blind.—Nevertheless, the wisdom of nature has implanted in us the predisposition to compassion in order to handle the reigns *provisionally*, until reason has achieved the necessary strength; that is to say, for the purpose of enlivening us, nature has added the incentive of a pathological (sensible) impulse to the moral incentives for the good, as a temporary surrogate of reason.[17]

Affect reveals, then, the providence of nature. The requirement for affect is a consequence of the empirical fact of the human constitution: namely, that human reason is wedded to something like a *life cycle*: conception, growth, maturation, decline. It has a natural course, just as the body does. The infant lacks reason entirely; the adolescent and very elderly sometimes have compromised reason. Nature has a problem that must be solved: with what do I *supplement* reason during its maturation or decline? Affect is the solution to that problem.

It is worth lingering on the teleology of affect, and in particular, the extent to which it deviates from Kant's vision in the "Essay." In the "Essay," the very essence of a malady of the mind is that it represents a dysfunction and aberration of the corresponding faculty. Kant presents this with the force of metaphysical necessity: in a world where there is no dysfunction, where there is no pathology, in which all of the faculties of the mind always carry out their functions, there could be no madness. Put theologically, had Adam and Eve and their descendants never sinned, there would be no madness, because madness is disease and there would be no disease. The *Anthropology* undercuts this doctrine, because even in such a world, there would still be the life cycle, the process of growth and maturation of our physical and mental powers; that is because the capacity to conceive and raise children was always part of God's original design: *go forth and multiply*. Thanks to affect, children would still harbor within themselves the motivation to do the right thing, even if they did not understand why it was the right thing to do.

Kant's attitude toward the teleology of *Leidenschaft* (passion) is more ambivalent. Kant says that, *unlike affect*, passion is inherently evil:

Passions are cancerous sores for pure practical reason, and for the most part they are incurable because the sick person does not want to be cured. . . .

That is why passions are not, like affects, merely *unfortunate* states of mind full of many ills, but are without exception *evil* as well.[18]

There is no good to come out of passion; it would appear to comprise no part of God's design for humanity. Passion is a malformation of *desire*. It takes place when an inclination or a tendency, say, a tendency toward vengeance, has consolidated itself, established itself, become crystallized as an overruling point of orientation for one's life. It is different from affect, which has a fleeting quality. Hamlet's desire for revenge is a passion; it forms the principle of his being. It has become constitutional: "Affect works like water that breaks through a dam; passion, like a river that digs itself deeper and deeper into its bed."[19] We can see, in passion, a representative of Burton's entire system of madness. Burton, of course, thinks that madness begins when one ruminates upon a distressing event for so long that the emotion it evokes—anger, grief, suspicion—becomes second nature.

It is perplexing, then, that after this vigorous condemnation of passion, *Kant rediscovers teleology inside of it*. The teleology of passion shows itself *not* when our faculties are in the process of development, as in affect, but in the process of decline. Being consumed with a passion can sometimes help a person to sustain and exercise his cognitive abilities when they would otherwise have atrophied. It can be a bulwark against the total deterioration of our cognitive powers. An example is an aging man who becomes convinced that he will win large sums of money in gambling:

> From time to time nature wants the stronger stimulations of passion in order to regenerate the activity of the human being, so that he does not lose the feeling of life completely in mere *enjoyment*. To this end it has very wisely and beneficially stimulated objects for the naturally lazy human being, which according to his imagination are real ends (ways of acquiring honor, control, and money).[20]

Unlike affect, which is clearly providential—it is part of God's design for our kind—passion is not clearly a matter of design. It is possible that the cognitive benefit of passion, far from being a matter of design, is a happy accident, an unintended but beneficial side effect of pathology. Kant leaves it unclear whether passion's role as a bulwark against cognitive atrophy is a design feature or a fortuitous accident. (Having flat feet can exempt you from a national draft; that is not a design feature but a fortuitous accident.)

The final appearance of teleology in Kant's system is by far the most subversive of the simple, original design of his early "Essay." For here, teleology insinuates itself into the heart of *Aberwitz*, the pinnacle of madness. For while the "Essay" depicts *Aberwitz* (there, *Wahnwitz*) as the ultimate desecration of reason, in the *Anthropology*, *Aberwitz* participates in the very nature of reason, and specifically in its *systematizing* tendency. Kant has rediscovered reason inside of madness.

Certainly, the bare idea that there could be reason inside of madness is an old one; Shakespeare's Polonius famously quipped, "Though this be madness, yet there is method in 't." One might read this, of course, merely as Polonius' implicitly discerning that Hamlet's madness is feigned; it is not true madness. But one could also read this as an observation about madness in toto. And certainly, it is part of Locke's system, whom we will return to shortly, that madness is marked by the preservation of one's inferential capacities: the mad, Locke declares, reason well from false premises. For Locke, reason *qua* intact inferential capacity is partly constitutive of madness; without reason, madness would not be madness; it would be *idiocy*. This theme, that madness must have reason inside of it, recurs throughout the nineteenth century, in Haslam, Wigan, and Pinel, as we shall see. What is baffling is that teleology appears here, in the *Kantian* system, which would seem, at least at the outset, to have no place for it.

In the *Anthropology*, madness recovers its voice; it speaks with the voice of reason, however muted:

> In [*Aberwitz*] there is not merely disorder and deviation from the rule of the use of reason, but also *positive unreason*; that is, another rule, a totally different standpoint into which the soul is transferred, so to speak, and from which it sees all objects differently. . . . It is astonishing, however, that the powers of the unhinged mind still arrange themselves into a system, and that nature even strives to bring a principle of unity into unreason, so that the faculty of thought does not remain idle. Although it is not working objectively toward true cognition of things, it is still at work subjectively, for the purpose of animal life.[21]

Because madness tends to systematize, we cannot see it as a pure or decisive departure from reason. It is a departure that carries a token of reason with it; it bears the stamp of its origin in the reasoning mind. And this, the systematizing quality of madness, is not an accident; it is by design; it is a special-purpose design feature, not a mere rudiment of reason. *Nature itself* has bestowed upon madness a systematizing quality, "so that the faculty of thought does not remain idle."

It is worth pausing here to appreciate the way in which teleology and dysteleology collide in Kant's system, for we will encounter this same dynamic again and again. It is quite possible for any physical or psychological trait—the human hand, or eye, or memory—to be dysfunctional in some respects and functional in others. It is rare that a biological organ is dysfunctional *simpliciter*. Consider the common flu, initiated by a dysfunction: a virus breaches the cell wall and hijacks the cell's protein-making ability in order to replicate itself. This triggers a cascade of biological reactions; some are further dysfunctions, and some are functional and adaptive responses to the infiltration. The fever that accompanies flu is a functional and adaptive response; it is not itself a dysfunction, but seems to have the purpose of helping to destroy the virus, though it is not yet entirely clear how it does so.

It is in precisely this manner that Kant discovers a fragment of purpose and goal-directedness inside *Aberwitz*. True, *Aberwitz* is, in its essence, a dysfunction of the faculty of reason: in this respect the Essay and the *Anthropology* are in perfect accord. But the fact that defective reason still attempts to systematize its mad productions represents one respect in which it shows a goal-directed character. Its goal, as Kant tells us, is to protect reason from total deterioration; it has a preserving function, it ensures "that the faculty of thought does not remain idle." That Kant recognizes that *Affect*, *Leidenschaft*, and *Aberwitz* have a teleological aspect or quality does not transform him into a "proponent" of *madness-as-strategy*. These two tendencies of thought, madness-as-strategy and madness-as-dysfunction, can often coexist in one and the same thinker, with one or the other predominating. The exact same dynamic repeats itself in Griesinger, who thinks that, though mental disorders are, in the main, organic dysfunctions, they may yet have some functional or goal-directed aspects. For example, while delusions stem from dysfunctions of the brain, their precise contents tend to be explicable as coping mechanisms, as "wish fulfillments."

Theorists in the early nineteenth century will continue to wrestle with this principle, that reason is buried inside of madness. In particular, we will see that both Haslam and Wigan react to this discovery, but in different ways. For Haslam, the appearance of reason inside of madness is, in a sense, the ultimate perversion of madness; it lends a distinctive quality of insidiousness, even evil, to madness. The essence of madness is to dissimulate itself, to feign reasonableness. However, madness can only feign reasonableness on account of there being reasonableness *inside* of it. The moral perversion of madness is this: madness rejects reason while utilizing it as a means of furthering its own persistence. When dealing with a madman, we must find a way to puncture the semblance of reason to reveal the madness inside of him, that is, to expose his madness *as* madness. Wigan takes a more optimistic approach. The fact that madness has reason inside of it is the cornerstone of therapy and healing. For the entire art of healing consists in locating that remnant of reason that persists inside of madness, and addressing it directly. When the wise physician is bombarded by the madman's accusations, his theatrics, his pronouncements, he takes the following stance: *I'm not going to listen to you; I'm not going to engage with you. I'm going to seek out the remnant of reason inside of you, that remnant that is still able to retain the slightest thread of doubt about the veracity of its mad productions; I will address myself directly to that remnant, for the sole purpose of fortifying it in its war against madness.*

IV.

Finally, we can engage, even fleetingly, with Locke. For even though Locke comes before Kant, it becomes far easier to grasp the outlines of Locke's conception of madness once we have understood Kant's. That is because we are now able to appreciate the extent to which Locke's vision of madness falls outside of this dysfunction-centered framework. Locke's vision is ultimately *anti-Kantian.* That is because, for Locke, whatever madness is, it does not consist in the breakdown or the disruption of a faculty of knowledge; *idiocy* does. We must seek the root of madness elsewhere.

Locke's most important discussion of madness appears in Chapter 11, Book 2, of *An Essay Concerning Human Understanding.*[22] At this point, he has just finished enumerating the faculties of the human mind. These are the faculties that God has equipped us with in order to know the world around us, and ultimately to know *him,* his existence and nature. At this point, Locke has identified six faculties (or seven, depending on how you wish to count them): perception, retention, discerning, comparing, compounding, abstraction. These are operations that the mind performs on ideas; they are things the mind can do with ideas.

Now we might expect—had Locke and Kant been cut from the same cloth—that Locke, when he arrives at the topic of madness, would tell us that there are exactly *six* different ways that one could be mad, because there are six different faculties of the mind. He does *not*, however, explain madness this way. Rather, he explains *idiocy* this way:

> How far *Idiots* are concerned in the want or weakness of any, or all of the foregoing Faculties, an exact observation of their several ways of faltering, would no doubt discover. For those who either perceive but dully, or retain the *Ideas* that come into their Minds but ill, who cannot readily excite or compound them, will have little matter to think on. . . . And indeed, any of the forementioned Faculties, if wanting, or out of order, produce suitable defects in Men's Understandings and Knowledge. In fine, the defect in *Naturals* seems to proceed from want of quickness, activity, and motion, in the intellectual Faculties, whereby they are deprived of Reason.[23]

Instead of subsuming madness, too, under this methodological principle ("discover which faculty is wanting or weak"), he exempts it. Immediately following this passage, he tells us: "Whereas *mad Men*, on the other side, seem to suffer by the other Extreme."[24] The other extreme of what? He does not say. But on the surface, the most natural reading is that madness is the opposite extreme of a "want of quickness, activity, and motion" in a faculty—which seems to imply that madness *involves the normal or even enhanced operation of one's faculties*. But then we have quit the Kantian framework, wherein madness results from a failure or aberration of a mental faculty. Madness, it appears, has an entirely different cause.

This reading—that, for Locke, madness is *not* a want or weakness in a faculty—is confirmed by the next passage: just after telling us that madman represents the *other extreme* of the idiot, he announces his celebrated formula:

> For [madmen] do not appear to me to have lost the Faculty of Reasoning: but having joined together some *Ideas* very wrongly, they mistake them for Truths, and they err as Men do, that argue right from wrong Principles. For by the violence of their Imaginations, having taken their Fancies for Realities, they make right deductions from them. Thus you shall find a distracted Man fancying himself a King, with a right inference, require suitable Attendance, Respect, and Obedience: Other who have thought themselves made of Glass, have used the caution necessary to presence such brittle Bodies.[25]

In madness, the reasoning ability, *qua* inferential capacity, has been preserved completely intact, as in Kant's *Verrückung*. The madman can correctly deduce propositions from other propositions; the problem is that his premises are false. Borrowing a metaphor from Cecilia Heyes, if we think of the ideas as the grist of thought, and the faculties as the mills of thought, then, in the madman, the *mills* of thought are serving their proper functions, but the *grist* of thought is somehow malformed.[26] Even better, we can draw a metaphor from computer science: *garbage in, garbage out*.[27] But then, *whence* these malformed ideas? Whence this ideational "garbage"?

In Book II, Chapter 33, we divine the root of madness: *association*. Association describes a tendency of the mind to glue together ideas irrespectively of whether they have a natural connection with one another.[28] It is not that association fuses ideas that *lack* a corresponding natural connection; it is that it fuses ideas *without regard* for such a connection. There are many ways such associations can be formed. They can be formed by education; for example, the ideas of goblins and sprites have no natural connection to darkness, but let "a foolish Maid inculcate these often on the Mind of a Child, and raise them there together, possibly he shall never be able to separate them again so long as he lives."[29] Associations can be formed by trauma; for example, I see a dead body in a room and from that point on I "can as little bear the one as the other."[30] Associations can be formed by rumination; they arise when I ruminate for so long on a wrong that was carried out against me that I cannot think of the one who wronged me without, at the same time, thinking of the wrong; my opponent and his deed are irrevocably fused into a single concept.

There are two positions in the scholarly literature on Locke's notion of association. One holds that association is, in and of itself, a "normal," "healthy," "functional" process; I use it correctly when I only fuse ideas that have a natural connection, and I use it incorrectly, even pathologically, when I fuse ideas that have no natural connection, and the latter is the wellspring of madness. Another is that association is, *in and of itself*, a pathological process for forming ideas; it is a contradiction in terms to say that I have "correctly associated two ideas." Association does not have a standard of correctness, any more than cancer does. In this view, association only comes into play when reason has failed in its task of assessing the natural connection between ideas. Reason alone should cement two ideas together when it has discerned a natural connection between them; it neither needs, nor should it rely on, association to carry out its proper job.[31] We need not enter into that debate, though, to establish our fundamental point: *association does not represent the failure of a faculty to perform its function in a Kantian manner, that is, in the manner of a breakdown or disruption.* At worst, it represents a misuse of one's reason in a manner that is contrary to how God intended it to be used. Put differently, it represents an abdication of the responsibility that attaches to being a person with a reasoning mind.

We must insist here, rather sharply, as we did in Burton, on the distinction between *misuse* and *defect*. The difference between the two is inherent in the very notion of function itself. Once we have the idea of an organ, faculty, or tool that has a function, we can discriminate, at least conceptually, between a breakdown in that function—for example, a defect—that amounts to a constitutional inability of that object to carry out its function, and an instance of misuse, in which the item (artifact, organ) is deployed in a manner it was not designed or intended for. Nobody would question that Ajax's impaling himself on his own sword represents a *misuse* of the sword rather than a malfunction or a defect in it. And the theological significance of this distinction is beyond question: if I shoot a gun and it explodes in my face, that is the fault of the manufacturer; if I point the gun at myself and pull the trigger, the damage is my own fault. Locke's system is still, like Burton's, and like Cheyne's, fundamentally wedded to the category of misuse; this is why Kant's system represents such a sharp break from it. Kant himself is rather emphatic on this point:

> I can not even in any way convince myself that the disturbance of the mind originates from pride, love, too much reflection, and who knows what misuse of the powers of the soul, as is generally believed. This judgment, which makes of his misfortune a reason for scornful reproaches to the diseased person, is very unkind and is occasioned by a common mistake according to which one tends to confuse cause and effect.[32]

At the risk of crudely oversimplifying, we might venture to suggest that Locke's writing expresses the closure of a certain era, an era in which madness is thought in terms of misuse; Kant's writing helps to open a new one, in which madness is thought in terms of dysfunction. Or, better—since the dysfunction-centered view stretches back to the Hippocratic authors—we can say that Kant's writings on madness mark the beginning of an era in which the concept of dysfunction, as an organizing principle for the study of madness, is awakened from a long, troubled, sleep.

6

The Mountebanks of the Mind

John Haslam, apothecary to Bedlam at the beginning of the nineteenth century, advances Kant's early, dysfunction-centered conception of madness. Haslam, however, openly wrestles with a question that did not occur to Kant: if the madman, in order to evade institutionalization, feigns reason, then doesn't that, ipso facto, make him reasonable, not mad? Put differently, what precisely is the relation between reason and madness such that the mad person can feign reason but not *be* reasonable (Section I)? This problem was particularly urgent for Haslam because of an "extraordinary difference of medical opinion" regarding whether a certain patient at Bedlam, one James Tully Matthews, was actually insane (Section II). The answer that Haslam eventually arrives at is that the mad person *understands, but rejects,* the claims of reason. For the mad person, reason lacks the *compellingness* that it possesses for the sane person (Section III). Haslam's solution sets up the basic problem of psychiatry: the physician must *outwit* the madman. With this formula, Haslam discovers a new feature of madness, a new telos: dissimulation. The goal of madness is to dissimulate reason in order to perpetuate its own existence *as* madness (Section IV).

I.

John Haslam, apothecary to Bedlam hospital from 1795 to 1816, fully accepts Kant's madness-as-dysfunction picture and formulates it in a new way. As we will see, however, he adds to this picture a new characteristic, a new quality, an essential feature of madness, which is *dissimulation*. Haslam was living at the beginning of what historians now think of as the era of mass institutionalization; the number of asylum patients skyrocketed during the nineteenth century. In England, the number of patients ballooned from just several thousand in 1800 to 100,000 in 1900.[1] The mad person is no longer the charge of the family or the community, but of the state. But this sets up a very basic

Madness. Justin Garson, Oxford University Press. © Oxford University Press 2022.
DOI: 10.1093/oso/9780197613832.003.0007

problem for the mad, who, as a rule, do not *want* to be confined. To protect against the threat of confinement, the mad must feign reasonableness.

But this raises a paradox, one that Haslam is sharply aware of. If the mad person can feign reason, does that not, ipso facto, make him reasonable, and hence, not mad? My ability to maintain, under multiple and extensive interviews, the facade of reasonableness, requires a firm grasp of my circumstances, the expectations and beliefs of other people, and the ability to manipulate, deftly, those expectations and beliefs—all capacities that we associate with reason, not with madness. So, this raises, in a very urgent form, a problem that Kant only hinted at: what, precisely, is the relation between reason and madness, if madness is not merely, as Kant discovered in the *Anthropology*, the opposite of reason, but if madness somehow carries reason inside of it or participates in the nature of reason? What, exactly, is the relationship that the mad person bears to reason such that he can participate in it while falling outside of the privileged circle of sanity?

Haslam begins his 1798 *Observations on Insanity* by outlining his basic ontology of madness, which is, at root, Kantian: the forms of madness are just so many ways that the faculties of the sound mind can be disrupted or fall out of working order.

> The sound mind seems to consist in a harmonized association of its different powers, and is so constituted, that a defect, in any one, produces irregularity, and, most commonly, derangement of the whole. The different forms therefore under which we see this disease [insanity], might not, perhaps, be improperly arranged according to the powers which are chiefly affected.[2]

What is intriguing is that this idea, that madness is simply a way the mind can err, was not, as it is today, so widespread that one could just state it and leave it at that, as if it needed no special argument or justification. Today, one can. Both the RDoC and *DSM* simply *assert* this principle: it requires no special explanation or defense; it is part of the accepted ontology of madness.

But Haslam *does* need such an argument. For madness-as-dysfunction is of sufficient novelty that intelligent and educated readers would demand a *reason* before nodding silently in approval. Fortunately, Haslam possesses not only a reason but a *foil*, one *Dr. Mead*, whom he positions as the interlocutor he must refute. Specifically, this Dr. Mead had dared to set forth, in print, the position that madness proceeds from *enhanced*, rather than *diminished* or *perverted*, mental powers. The madman, in Mead's view, suffers from a *superior* faculty of the imagination: for Mead, "this disease consists entirely in the strength of the imagination."[3] In response, Haslam argues that the madman cannot suffer from an augmentation of one of the mental faculties *because an augmentation in a faculty is a mark of excellence, even genius.* But the madman is not a misunderstood genius. Hence, madness affects the cognitive capacities not by way of an augmentation, but rather, by a diminution or perversion:

> The increased vigor of any mental faculty cannot constitute intellectual disease. If the memory of a person were so retentive, that he could reassemble the whole of what he had heard, read, and thought, such a man, even with a moderate understanding, would pass through life with reputation and utility. Suppose another to possess a judgment, so discriminating and correct, that he could ascertain precisely, the just weight of every argument; this man would be a splendid ornament to human society. Let the imagination of a third, create images and scenes, which mankind should ever view with rapture and astonishment, such a phenomenon would bring Shakespear to our recollection.[4]

This recognition, that madness necessarily involves a diminution, not an augmentation, of one's faculties, does not imply that the mad are incapable of extraordinary feats of poetry or philosophy or oration or invention. Rather, even if a particular person shows an augmentation in *certain* abilities, say, poetry or oration, this comes at the cost of a diminution or perversion of *other* faculties:

> Had Dr. Mead stated, that, together with this increased strength of the imagination, there existed an enfeebled state of the judgment, his definition would have been more correct. The strength, or increase of any power of the mind, cannot constitute a disease of it; strength of memory, has never been suspected to produce derangement of the intellect: neither is it conceived, that great vigor of judgment can operate in any such manner.[5]

If we are willing to concede that some madmen enjoy enhanced ability in one power or faculty, there must exist, alongside of it, a decreased, perverted, or abolished ability in another.

In addition to involving a decrease or perversion of a faculty, madness also has an organic basis. It stems from a brain disorder. So, Haslam's basic ontology is that the brain disorder causes the failure of a faculty, and the latter constitutes a form of madness. But this is not a simplistic psychoneural reductionism, for Haslam is adamant that social, psychological, and environmental factors can trigger the inner pathology that generates madness. In fact, for Haslam, the neurological basis of mental illness is arrived at not only through painstaking observation—Haslam carried out dozens of autopsies to investigate this problem—but also through an a priori demonstration. Because we are hybrid physical and spiritual beings, the disorder, the derangement, that gives rise to madness, must either reside in our spiritual substance or in our physical substance. But because our spiritual substance is incorruptible and eternal, then the disorder from which madness springs must reside in the *physical* substance:

> Is it not more just to conclude, that such organic affection has produced this incorrect association of ideas, than that a being, which is immaterial, incorruptible and immortal, should be subject to the gross and subordinate changes which matter necessarily undergoes?[6]

And while Haslam apologizes for engaging in what he calls a "metaphysical controversy," and while he lays out his position with due humility, we nonetheless see this line of reasoning recur repeatedly in nineteenth-century texts, such as Spurzheim, Gall, and Cox. But this theological motif does not undermine the scientific status of his investigation: science always begins with an a priori grasp of its domain.

II.

Haslam's doctrine is the Kantian one. But Haslam adds a new quality to madness: *dissimulation*. To appreciate this, we must see how the theme of *fraudulence* permeates Haslam's corpus. It is no exaggeration to say that Haslam is obsessed with fraudulence: how to detect it and how to expose it. The key text here is his *Illustrations of Madness* of 1810.

Illustrations of Madness, at first glance, appears to be a report on a particular case of madness, that of James Tully Matthews. Matthews was involuntarily committed to Bedlam in 1797; immediately thereafter, his family members began to plead—unsuccessfully—for his release on the grounds that he was not, in fact, insane, but that he merely held "philosophical" beliefs of a somewhat bizarre character. Twelve years later, in 1809, Matthews' family commissioned two doctors to interview Matthews—one Dr. Henry Clutterbuck and one Dr. George Birkbeck—and after extensive examination, both doctors concluded that Matthews was "perfectly sane." In response to this professional affront, Haslam gathered all of the top physicians in London to interview Matthews. They penned a deposition that Matthews was, in fact, "in a most deranged state of intellect, and wholly unfit to be at large."[7] Matthews remained in confinement until his death in 1815.

Illustrations of Madness is a text that is almost impossible to classify. Equal parts science and satire, the full title reads: *Illustrations of Madness: Exhibiting a Singular Case of Insanity, and a no less Remarkable Difference in Medical Opinion: Developing the Nature of Assailment, and the Manner of Working Events; with a Description of the Tortures Experienced by Bomb-bursting, Lobster-cracking, and Lengthening the Brain*. The title alone indicates that Haslam is attempting to accomplish several goals at once. First and foremost, the book is a contribution to science through the meticulous observation of a *Singular Case of Insanity*. Second, it is a contribution to the sociology of medicine in that it offers a detailed account of a *Remarkable Difference in Medical Opinion*. Third, and finally, it promises a lurid description of almost incomprehensibly cruel tortures, including *Bomb-bursting, Lobster-cracking, and Lengthening the Brain*. This is puzzling. These are, of course, the names of various assailments that Matthews alleges are being inflicted on him by a mysterious gang. Haslam, in the title, pretends that these afflictions are real and that they are, in fact, being practiced on Matthews daily. So, the title alone carries a touch of sarcasm, a smile, a wink, to the perceptive reader. But this raises a profound question that it will take some time to answer: why is this touch of sarcasm, the smile, the wink, appropriate for a putatively "scientific" text? Answering that question will ultimately allow us to grasp the outlines of Haslam's thought.

Turn now to the title page. It contains a short epigraph, drawn from Samuel Foote's 1794 satire, *The Devil upon Two Sticks*: "Oh! Sir, there are, in this town, Mountebanks for the mind, as well as the body." What does this mean? How are we to read this?

To understand the epigraph, it will help us to consider the context of the passage. In the scene from which the epigraph is drawn, the devil has approached the protagonist, Invoice, and is recommending various forms of medical charlatanism, which Invoice refuses to practice:

Devil: Oh, Sir, I don't design to engage you in any personal service; I would only recommend it to you to be the vender of some of those infallible remedies, with which our newspapers are constantly crouded?
Inv: You know, Sir, I am possessed of no secret.
Devil: Nor they either: A few simple waters, dignified with titles that catch, no matter how wild and absurd, will effectually answer your purpose: As, let me see now! Tincture of Tinder, Essence of Eggshell, or Balsam of Broomstick.
Inv: You must excuse me, Sir; I can never submit.[8]

The devil then tempts Invoice with a different *kind* of charlatanism: a charlatanism *not of the body, but of the mind.*

Devil: I think you are rather too squeamish. What say you, then, to a little spiritual quackery?
Inv: Spiritual?
Devil: Oh, Sir, there are in this town mountebanks for the mind, as well as the body. How should you like mounting a cart on a common, and becoming a Methodist Preacher?[9]

The first question that we must address is this: according to Haslam, *who are these mountebanks of the mind*? The obvious answer to this question is Drs. Clutterbuck and Birkbeck. Put simply, they are frauds. Haslam goes to great lengths to explain how, just because one earns a degree in medicine, that does not, in and of itself, make one competent to practice medicine. For what if the institution that granted you a medical degree is a fraudulent institution? Then you would be a fraudulent physician. As he puts it:

Every person who takes the degree of Doctor becomes, in consequence of taking such degree, a learned man; and it is libellous to pronounce him ignorant. It is true, a Doctor may be blind, deaf and dumb, stupid or mad, but still his Diploma shields him from the imputation of ignorance.[10]

But this *mountebank* also refers, *equally justly*, to somebody else: James Tully Matthews himself. For Matthews has spent dozens, if not hundreds, of hours dissimulating reason. He is a con man, a swindler, a counterfeit, in matters of the mind. It is worth pausing to contemplate the exceptional situation that Haslam finds himself in. He is in a situation where two frauds are being perpetuated in parallel: people who claim to be doctors, but really are not, are interviewing someone who claims to be sane, but really is mad. Haslam is watching people who fraudulently claim the title of doctor interacting with a man who fraudulently claims the title of sanity. *Surrounded by fraudulence*, surrounded by mountebanks of the mind, what is a reasonable person to do, but to playfully dissimulate, to playfully enter this game of make-believe, to perpetuate his own little act of fraudulence?

Haslam's miniature act of fraudulence is twofold. On the one hand, he pretends to believe that Matthews, a madman, is really sane; on the other, he playfully treats Clutterbuck and Birkbeck as if they were legitimate doctors, *as if they were his colleagues*. Part of the pleasure of the text lies in the way that Haslam instigates his own fraudulence. And lest the reader get confused about the nature of the text, Haslam apologizes in advance for his playful tone: "If, contrary to his expectation, the Reader, throughout this narrative, should suspect a sneer, the benevolence of the Writer allows him to soften and correct it by a smile."[11]

Haslam dispenses with the first group of frauds, Drs. Clutterbuck and Birkbeck, very rapidly. Haslam begins by referring to them *as if they were* his esteemed colleagues, and pretends that perhaps they, in their superior medical wisdom, have discerned something in Matthews that he and his colleagues at Bedlam have systematically overlooked:

> It may here be proper to state that it had been the unvarying opinion of the medical officers of Bethlem Hospital, that Mr. Matthews had been insane from the period of admission to the present time. Such opinion was not the result of casual investigation; but a conclusion deduced from daily observation during thirteen years. But aware of the fallibility of human judgment, and suspecting that copious experience which sheds the blessings of light upon others, might have kept them in the dark. . . [12]

But Haslam cannot maintain the façade for very long before he says what he really thinks: "It should be recollected that these gentlemen have much practical experience, and are competent judges of all systems of error but their own."[13]

The second target of his missive is Matthews. (Of course, to obtain the mad ramblings on which *Illustrations of Madness* is based, he had to engage in a bit of dissimulation himself: he had to convince Matthews to produce the text for him at his request. These were not papers he found under Matthews' mattress, or notes feverishly scribbled on toilet paper; these were documents that he requested of Matthews, who was, Haslam assures us, "contented and grateful" to have the opportunity to introduce "his philosophic opinions to the notice of a discerning public." No doubt Matthews produced these notes with the hope that, when the texts were published, they would warn others similarly afflicted.) Matthews' text is an expose of a gang. Here are their characteristic workings:

> [Matthews] insists that in some apartment near London Wall, there is a gang of villains profoundly skilled in Pneumatic Chemistry, who assail him by means of an Air Loom. . . . The assailing gang consi[s]ts of seven members, four of whom are men and three women. . . . They leave home to correspond with others of their profession; hire themselves out as spies, and discover the secrets of government to the enemy, or confederate to work events of the most atrocious nature. At home they lie together in promiscuous intercourse and filthy community.[14]

An irony of this text is that Matthews was, in fact, arrested by the French and imprisoned for three years, between 1793 and 1796. He had sought to broker peaceful relations between France and England and was arrested on suspicion of spying.[15] So, it is not entirely a stretch of reason for Matthews to think that someone might wish to use advanced technologies to extract secrets from him.

Haslam engages with Matthews' text as if it were an important scientific document, as if Matthews and he were fellow scientists:

> Whoever peruses a work on Nosology will be painfully impressed with its formidable catalogue of human miseries; it therefore becomes exceeding distressing to swell the volume with a list of calamities hitherto unheard of, and for which no remedy has been yet discovered.[16]

After being transferred to another institution, Matthews died in 1815; the following year, Haslam was forced to resign from Bedlam because of complaints about his treatment of inmates, *and partly on the basis of a document written by Matthews himself accusing him of malpractice.* One can only imagine Haslam's bewilderment and fury to be forced to defend his behavior from the written accusations of a lunatic. In regard to the accusatory documents produced by Matthews, Haslam said in his defense:

> I conceived that its circulation [that is, of Matthews' accusation of malpractice] ought not to be prevented, on the presumption that there existed in the judgment of those who passed for persons of sound mind, a sufficient disrelish for absurdity, to enable them to discriminate the transactions of daylight, from the materials of a dream.[17]

Matthews' fraudulence only reached its zenith posthumously: he had succeeded in deceiving not only Drs. Clutterbuck and Birkbeck, *but the entire governance of Bedlam hospital.*

III.

Still, we have not yet answered our question: if the madman can feign reasonableness, even to the point of deceiving genuinely reasonable people, then he must, in some measure, be reasonable himself. But if he is reasonable, how can he be mad? Put more rigorously: what, precisely, is the relationship between the mad person and reason, such that the mad person can feign reason but still be mad? This question only appears to have arisen in stark form during the nineteenth century; before that, madness was, as Roy Porter put it, "publicly transparent;" madness was simply a matter of "behaving crazy, looking crazy, talking crazy."[18] Kant himself had grasped the paradox in a parenthetical remark in his early "Essay": "[In] the demented and insane persons, the understanding itself is attacked, it is not only foolish to reason with them (*because they would not be demented if they could grasp these rational arguments*), but it is also extremely detrimental."[19]

We must dive more deeply into this question. The madman deceives. This is what he does. Haslam has remarked on this throughout his entire career. In his 1798 book, *Observations on Insanity*, he presented the case of a mad person, identified only as "Case XXII." This was an "artful and designing man" who was so adept at deceiving, he attempted to convince his friends that the *physicians of Bedlam alone* have the right to render judgment about his mental health.[20] In other words, Case XXII was able to think something along the following lines: *I am going to trick this physician; I am going to pretend to be sane; the way that I will do so is by insisting that the physician alone has the right to judge my sanity; in fact, I will become indignant when my friends and family suggest that anyone besides my physician has the authority to render such a weighty verdict; in this way, the physician will think me a reasonable man indeed, and perhaps release me from confinement.*

Moreover, this dissembling tendency is not the exception. *It is the norm*:

> They [the mad] have sometimes such a high degree of controul over their minds, that when they have any particular purpose to carry, they will affect to renounce those opinions which shall have been judged inconsistent: and it is well known that they have often dissembled their resentment, until a favourable opportunity has occurred of gratifying their revenge.[21]

The question, then, is this: if the mad person is so artful at feigning reasonableness, then how is it possible that she is not, in fact, reasonable, but mad?

We will approach the question from a somewhat different direction. A mad person says something reasonable ("only the doctor is qualified to render judgment about my mental health"). A sane person says something reasonable ("only the doctor is qualified to render judgment about my mental health"). What is the difference between the mad person and the sane person, such that the latter is *truly* reasonable but the former is not? It is, of course, this: *When the sane person says something reasonable, she is just being herself; she is expressing her intrinsic reasonableness. When the mad person says something reasonable, she is not being herself: she is feigning reason.*

But this raises a final question, and with this, our journey reaches its end: how can the madman "play" at being reasonable, *without being swayed by the inherent reasonableness of what he says*? How can the madman understand and even articulate, eloquently, a cogent trail of reasoning, without being persuaded by his own words? Ultimately, the difference between a mad person and a sane person comes down to this: a sane person is *compelled* by reason; a mad person is not.

Perhaps the best way to understand how a person could *fail to be compelled* by reason is by contemplating the psychopath. The psychopath, we are often told, appears to grasp, *and even assent to*, moral propositions such as "It is immoral to hurt an animal for fun." Her deceptiveness consists in this: she can say things like that; she even understands these statements—or at least she understands that society believes in these statements even if she herself does not—but *they do not compel her*; they have no motivational force upon her mind and heart. This is a puzzle that philosophers and psychiatrists alike have wrestled with: *is* it possible, and if so, *how* is it possible, that the psychopath understands the claims of morality without being swayed by them? And there are roughly two paths, two positions, that one may adopt on the problem.[22] The first is that the psychopath *does* understand, perfectly well, that, say, it is wrong to torture an animal for fun, but this understanding is simply not accompanied by any motivational power; her belief simply fails to engage with the machinery of motivation or behavior. There is no question of a *cognitive* shortcoming; this is a failure of the will. Even so, it is hard to accept that someone can understand, *fully and completely*, that it is wrong to torture an animal for fun but have no motivation whatsoever to act upon that knowledge. In other words, it seems, at least to some theorists, that it is impossible to truly understand a moral rule without having *some* motivation, however feeble, to act on it. So, according to this second point of view, *the psychopath does not really understand moral statements*. What the psychopath understands, when she hears and even assents to statements like "It is wrong to torture an animal for fun," is that *society disapproves of torturing animals for fun*, but she understands that in the exact same way she understands that society disapproves of wearing a sleep gown in public, or not removing your hat during the national anthem: it marks nothing more than an arbitrary social prejudice. Here, Haslam grapples with a parallel puzzle: how can a madman understand, and even articulate, a perfectly reasonable piece of oration, but fail to be compelled by it?

A hint to the solution can be found in Lewis Carroll's celebrated 1895 puzzle, "What the Hare Said to the Tortoise," popularized by Douglas Hofstadter.[23] Achilles gives Tortoise a perfectly valid argument:

(A) Things that are equal to the same are equal to each other.
(B) The two sides of this Triangle are things that are equal to the same.
Therefore,
(Z) The two sides of this Triangle are equal to each other.

Tortoise wants to know the basis of the inference, that is, why he should accept Z on the basis of A and B. Achilles responds with another premise:

(A) Things that are equal to the same are equal to each other.

(B) The two sides of this Triangle are things that are equal to the same.

(C) If A and B are true, Z must be true.

Therefore,

(Z) The two sides of this Triangle are equal to each other.

Achilles, triumphant, tells Tortoise that *if* he accepts A, B, and C, then he *must* accept Z. Tortoise does not see why he *must* accept Z on the basis of A, B, and C. Achilles responds by asserting that *if* A and B and C are true, then Z *must* be true. Tortoise responds, of course, by asking Achilles to add the latter statement as another premise to the growing argument, to see if the argument becomes more persuasive. Our premises, now, include the following:

(A) Things that are equal to the same are equal to each other.

(B) The two sides of this Triangle are things that are equal to the same.

(C) If A and B are true, Z must be true.

(D) If A and B and C are true, Z must be true.

(Z) The two sides of this Triangle are equal to each other.

Tortoise is a model of someone who appears to understand logic, *without being compelled by it*. The *compellingness* of logic is absent. Now, for those of us who both understand the argument and are compelled by it, that is, for those of us who were compelled the first time around, it is as if there is a natural transition, a clicking-into-place, an inner *mechanism* that operates in the following fashion: when you have these two premises as beliefs, *A*, and *if A then B*, then create a new belief *B*. So, it is easy to surmise that in the madman, there is an inner mechanism that is *supposed* to generate new beliefs according to modus ponens. One's inability to feel any compulsion to accept the conclusion of a sound deductive argument is explained by a mechanism that fails to do its job.

We have finally discovered the relationship between the madman and reason: the madman is somebody who understands the claims of reason, and yet has rejected them. His relationship to reason is one of rejection. Like the sociopath, he understands those claims (or at least understands that the *others* feel compelled by such reasoning): he understands them well enough to be able to appropriate them, to use them as further grist for his project of dissimulation.

IV.

Hence the new character, the new figure, the new quality of madness that Haslam discovers: dissimulation. This is the essence of madness: madness, by nature, dissimulates reason without accepting the claims of reason, without being compelled by them. If we wish, we can think of this as a kind of telos of madness. The goal of madness is to dissimulate reason in such a way as to evade detection. Madness has the being of a parasite like *Trypanosoma brucei*, which regularly changes its protein coat to evade detection by the host.

The dissimulating character of madness sets up a new goal for the psychiatrist: the psychiatrist must constantly monitor the madman *in order to outwit him*. Haslam goes to great lengths to instruct his colleagues—the authentic colleagues, not the fraudulent ones—in the art of outwitting the madman. And this is where the difference between madness and reason yields its clinical application. The madman is fundamentally not reasonable. He has rejected the claims of reason and at this point can only simulate reasonableness. And this is the principle for outwitting him: *the madman can sustain this simulacrum for only so long before he tires of the ruse, before he betrays his inner madness*:

> Unthinking people are frequently led to conclude that, if during a conversation of a few minutes, a person under confinement shall betray nothing absurd or incorrect, he is well, and often remonstrate on the injustice of secluding him from the world. . . . In the same manner, insane people will often, for a short time, conduct themselves, both in conversation and behavior, with such propriety, that they appear to have the just exercise and direction of their faculties; but let the examiner protract the discourse, until the favourite subject shall have got afloat in the madman's brain, and he will be convinced of the hastiness of his decision.[24]

Consider a sane person. A sane person is a reasonable person; that is simply her nature. Acting reasonably does not require any special strain or energy because it is just a matter of being herself. For the madman, acting reasonably requires putting on a role; it requires being someone he is not. It demands tremendous exertion. There are thus two keys that are requisite in this project of exposing the madman's unreason: time and trust. You need time with him. You discuss topics with him that are not pertinent to the illness; you touch on an area related to the illness; you back away; you gently draw the madness to the surface. You also must become his confidant; you earn his trust, by conveying that perhaps you are, in fact, open to receiving his innermost secret. Perhaps you, too, have heard rumors of the Air Loom Gang, or perhaps you yourself have long suspected that the government uses advanced technologies to extract secrets from unwilling or unwitting victims. Perhaps you are even one of the allies in his strange quest. Sooner or later he will *let down his guard*. He will betray his secret. And this is another reason and justification for institutionalization. Not only does it keep society safe from those who are unfit to be at large, not only does it have therapeutic value in removing a person from a pathogenic environment, but you need it in order to cultivate a relationship and have the time to converse at length with the madman and thereby lead him to betray his madness.

We can now understand something that may have, at first, appeared incidental to Haslam's practice, namely, the autopsy. Haslam performed dozens of autopsies in search of a physical malformation or defect underlying madness, and though he never found one, he did find various and sundry neurological abnormalities in the brain. In short, Haslam was in search of a *biomarker*, some kind of biological structure or foolproof signature that reveals madness "directly," immediately, without the need to sift through ambiguous impressions and dissimulating presentations. Autopsy, he hoped, would exhibit the truth of madness in a decisive way, in a way that the clinical interview simply cannot.

Wigan, too, like Haslam and Kant, discovers reason inside of madness; he finds that madness somehow participates in reason. For Wigan, this is not only part of its essence, but *a crucial part of its treatment*: in the best of cases, one can cure madness—or at least draw someone who is on the precipice of madness firmly back into the realm of sanity—by appealing to his reason. The fact that the mad are capable of reasoning, the fact that they have some remnant of reasoning power inside of them, is the key to successful therapy.

Wigan realizes, as does Haslam, that if someone is firmly in the grip of delusion, there is no point trying to have a frank discussion with her about her delusions. Wigan is by no means laboring under the misguided assumption that a mad person can be "talked out of her madness." Rather, one intervenes *before* the delusions become solidified, during that fragile transitional state in which the patient experiences a profound uncertainty; the delusional ideas have presented themselves to her judgment merely as worthy of being entertained, merely as having some initial plausibility, but she has not yet taken the leap; she has not yet made a decision for or against them; she is "on the fence." This is the moment at which the physician might be able to intervene by appealing to her reason, and thereby *fortifying* her reason in its struggle against madness. Wigan, like Haslam, sees in madness a dissimulating tendency, but it is not dissimulating in the sense that it attempts to deceive *the other person*. It is dissimulating in that it attempts to deceive, to misdirect, to subvert, the patient's *own* reason. And this is possible because *each of us is many*. Inside of us there is more than one mind, and the multiple minds can engage with one another, interact with one another, hoodwink one another, in much the same way that numerically distinct persons can.

So, both Wigan and Haslam recognize that the mad person *somehow or another* participates in reason. (For Wigan, this is precisely what distinguishes madness from idiocy.) For Haslam, the mad person only participates in reason in the manner of playing a game or adopting a strategy for obtaining a goal: being reasonable is antithetical to his nature, not because he does not understand the claims of reason but because, deep inside of him, he has rejected those claims. Being reasonable is merely a ruse; it is a role that is imposed upon him by the demands of the interpersonal world. Madness differs from idiocy in that, while idiocy is the absence of reason, the madman lives in a state that *opposes* reason. *There is nothing left to do but to expose him for what he is.* Haslam's position on the possibility of treatment is a fundamentally pessimistic and even cynical one, despite his occasional claims to the contrary. For Wigan, in contrast, the madman participates in reason because *he himself is dual*. There is, inside of him, a *remnant of reason* that courageously battles the forces of madness that nonetheless threaten to close in on all sides, that threaten to choke off that remnant. But the remnant is also a seed, and this is the ground of our hopefulness: that, no matter how small it is, that seed might, with time and care, germinate and, with it, return him to a state of full health.

7

The Miracle of Sanity

Arthur Ladbroke Wigan, in his *A New View of Insanity: The Duality of the Mind* of 1844, rethought the possibility of madness not as a defect or injury to the mind, but as somehow inherent in its very structure. That is because, for Wigan, each of us has *two* minds, corresponding to the two hemispheres of the brain. Madness happens when the two minds pursue radically different courses (Section I). More specifically, madness occurs when one hemisphere is sick and the other hemisphere is healthy. Madness requires a kind of relationship or dialogue between the sick mind and the healthy mind: the sick mind must gain some ascendency over the healthy one (Section II). This division creates a new system for classifying the mad, in that *there are as many forms of madness as there are ways that the healthy mind relates itself to the sick mind*: the healthy mind can *rebuke* the sick, it can be *deceived* by the sick, and it can *acquiesce* to the sick. But then all forms of madness are infused with teleology: in the first, reason strives and fails to conquer madness; in the second, madness tricks reason; in the third, reason accepts madness as a contrivance for coping with a painful reality (Section III). Wigan, at long last, comes to reveal the function of this dual structure: having two minds marks the difference between human and beast, reason and instinct, reflection and spontaneity (Section IV). Finally, Wigan confronts the possibility that reason can sometimes *knowingly choose madness over reason* (Section V).

I.

Wigan, in his *A New View of Insanity: The Duality of the Mind* of 1844, was one of the first physicians to rethink the possibility of madness neither as a surge of divine wrath, nor as a contingent accident, something that befalls one, like plague or injury, but as buried in the structure of reason itself. The possibility of madness, for Wigan, is an a priori condition of reason. It is nestled inside of reason. There could not be a form of reason, at least a *human*

Madness. Justin Garson, Oxford University Press. © Oxford University Press 2022.
DOI: 10.1093/oso/9780197613832.003.0008

form, that is not on the precipice of madness. Reason exists only through a perpetual struggle against madness. It requires effort, exertion, vigilance, attention, willpower *not* to be mad.

In a sense, Wigan's achievement in psychiatry is similar in kind to Isaac Newton's in physics. Newton's profound accomplishment was to reformulate the basic question of physics. With Newton, the question was no longer: *why do things move?* The basic question was: *why do they ever change the way they move?* Simply by shifting the question, Newton broke decisively from Aristotle and brought physics into the modern world. Similarly, for Wigan, the basic question of psychiatry is not: *how is madness possible?* The question is: *how is sanity possible? How is anybody sane? How is it possible to be sane, and stay sane?* And this question cannot be understood in the trifling sense of how one can remain sane, *in the face of* the tragedy of everyday life, in the face of one's painful formative environment, in the face of hurricanes and wars, oppression and small-mindedness, looming ecological catastrophe, etc. etc. The question must be understood as: *how is anybody sane, given the dual structure of reason?* Sanity requires a constant and even heroic striving. Madness is constantly pounding at the door.

Moreover, the way in which the possibility of madness is implicit in the structure of reason cannot be identified with the way in which disease, generally, is implicit in health. Surely, the possibility of having a heart attack is "implicit," in some sense or another, in having a heart. This is trivial. But this latter point—the possibility of heart attacks is implicit in having a heart— simply amounts to acknowledging that the possibility of failure always lurks in the performance of the proper function. Dysfunction surrounds function like a penumbra. But this is not at all the way madness relates to reason. Madness is not a dysfunction of reason but *a byproduct of its proper functioning*: madness *requires* a moderately functional, intact reasoning ability. The cessation of reason is not madness but idiocy or senility, as Locke proved. Madness must harness or co-opt reason; it must compel reason to serve its own perverse end. Madness harnesses reason the way that autoimmune diseases like HIV harness the body's immune system and force it to serve a malign end.

To understand how the possibility of madness is intrinsic to the structure of reason, and in what sense it is intrinsic, we must take a "detour" through the workings of the brain: we must first travel with Wigan into the realm of *established scientific fact*; we must leave behind all theorizing; we must part ways with all philosophical conjecture. Here are those facts. Let us follow

Wigan in his journey into the brain. Cut open the skull. Slice through and peel back the meninges. We find the cerebrum,

> a mass of tortuous folds like intestines, huddled together irregularly . . . the thick membrane which covers them is attached to the skull along its central line, from front to back, and dipping down between those convolutions, divides them into two equal and similar parts, called the *hemispheres* of the brain.[1]

The first thing that strikes us is the brain's duality, its "two-ness." The brain's two hemispheres are nestled, packed tightly into a small space, with some kind of wall or partition between them, the corpus callosum. Those are the facts.

Yet already, in our attempt to stick rigorously to the facts, we encounter a puzzle of a philosophical nature. We just said that in the skull we find *one* brain with *two* hemispheres. But this is peculiar. We do not say, "In the head there is *one* eye with *two* lobes." We say there are two eyes. We do not say, "A person has one kidney with two lobes." A person normally has two kidneys. So, why do we say that a person has one brain with two hemispheres, rather than two brains? The reason, he thinks, is the utter failure, on the part of almost all thinkers who have studied the mind, to recognize that "the mind is essentially dual, like the organs by which it is exercised."[2] We cannot, at this point, explicate this *duality of the mind*. But we will accept it, for now, not as a clearly articulated doctrine, but as a placeholder, as a label, for a doctrine that Wigan's book aims to demonstrate irrefutably. *There are two brains, and each brain has its own mind*—so each person has two minds. Each of our two brains has its own ability to think, feel, will. But in ordinary life, when all is going well, the two brains are synchronized tolerably enough so that we do not experience a sense of discord or confusion or of being at odds with oneself.

But once we understand that each of us possesses two brains, that is, two minds, then suddenly, all of those phenomena, all of the ways of "being at odds with oneself," are given a simple and lucid explanation: they result from having two minds. Consider ambivalence. I am ambivalent, say, about whether I believe in God, or whether I want to have children. This happens when one mind embraces a proposition that the other mind rejects: "*One brain believed, and the other did not believe*."[3] Consider akrasia, weakness of the will: "Every man is, in his own person, conscious of two volitions, and very often conflicting volitions."[4] That is a consequence of having two brains: the one wants the sinful pleasure that the other detests. And finally, madness: madness is a consequence of having two brains, one healthy, one diseased, as will be explained.

Of course, the bare idea that the mind is somehow fractionated, divided, at odds with itself, was already foreshadowed by the phrenology of Gall and Spurzheim. In fact, one reason for the lackluster reception of *A New View of Insanity* was that theorists did not see how Wigan's vision represented a departure from phrenology.[5] The crucial difference, however, was that phrenology was ultimately a faculty psychology; it localized the diverse personality traits and mental operations to different regions of cortex. This is a far cry from Wigan's view that each person is composed of *two distinct, whole minds,* each of which possesses the entire spectrum of mental competencies and personality traits. Moreover, unlike the phrenologists, Wigan never sought a more fine-grained form of localization than simply to assert that each hemisphere is a distinctive brain, and hence mind, in its own right.[6]

What is the function of having two brains? Why did God design us that way? Why were we *not* designed with a single brain? We cannot answer that question *fully* now. As Wigan's text goes on, as we follow him further in his wanderings, he adds more and more functions to this dual arrangement. But at least at the outset, and in a very preliminary way, we can say that a core function of this arrangement is what the biologists call "functional redundancy." The brain is an organ of vital importance not only physiologically, but spiritually, too. For the brain, as the organ of the mind, is what gives us the capacity to reason, to evaluate, and to choose, and, *along with choice,* confers moral agency upon us. I can be found praiseworthy or blameworthy for my actions. On the day of judgment, I can be found acceptable or unacceptable because of my choices. But the brain is fragile; it is susceptible to disease and senescence. Therefore, God has given me two brains, for the exact same reason he has given me two kidneys; the one can serve as a backup in case the other is compromised or destroyed:

> Ratiocination is so essential to the well-being of the individual, that the possession of two organs for this purpose, each capable of carrying on the function when its fellow is impaired or annihilated, seems only one more of the superabundant examples of design and contrivance in the structure of man, as a provision against accident or disease.[7]

This description of the brain's function, however, is problematic, for it suggests a certain picture of the brain that Wigan later takes some pains to refute. It suggests that there is no added benefit—no "cognitive benefit," as we would now say—from having two brains rather than one. Compare two people: the first is normal and has two brains; the second is abnormal and has only one. Now, suppose that the second person, the abnormal person, lives her entire life without ever experiencing brain disease. Then, according to this simplistic picture, she would do just as well—she would have the same "cognitive capacities"—as the person with two brains, just as a person with a single kidney that functions optimally throughout life would do "just as well" as a person who has both. But this is what Wigan comes to deny. Much later in his book, he tells us that the rationale for having two brains is not just functional redundancy; it is not just about having some spare parts lying around. Having two brains confers certain cognitive capacities that are *constitutive of human reasoning*. Having two brains explains the cognitive differences between human beings and other animals; it is, in fact, the difference between mind and instinct; it is the ability to reflect, for reflection just consists of *one brain observing what the other is thinking*. As a consequence, madness proceeds from the very dual structure that makes reason itself possible. It is in that way that the possibility of madness is prefigured in the structure of reason.

II.

On to madness proper. Let one of your two brains be attacked by an organic illness. Its associated nerves are damaged in such a manner that it begins to produce non-veridical sensations. That is, the brain's cortex is stimulated in exactly the way it would be if there were an object right in front of you; for example, it is being stimulated in just the way it would be were a dagger suspended in air right before your eyes. Now, the sick brain reports to its healthy neighbor that *there is a dagger*. The healthy brain, however, is anchored in reality, and by virtue of its being anchored in reality, it provides a check on the sick brain: *I had the weird thought that there is a dagger in front of me. But I can clearly see that there is no dagger. It must have been a flight of my imagination. Perhaps I'm just tired and I need some rest.*

All of us, Wigan thinks, are susceptible to this: you have a passing fancy or you see something out of the corner of your eye or you hear your name called as you're falling asleep, but you know it isn't real, and you *reject the evidence of your senses.*[8] The way this works is that one brain is mildly pathological and it is generating a false report, but fortunately, the second brain is correcting it.

What these examples demonstrate is that, once we have two brains, each brain has the job of exercising some *vigilance* over the other's reports. Each brain must recognize that the other can sometimes get things wrong, and it has to be willing to serve as a *reality check* for the other. In order to be a healthy person, in order to be "of sound mind," you need to be on the lookout; you need to be cautious; you need to exercise discretion. Being "of sound mind" is not something that comes for free; it is not something that comes with being a normal human being. It is the result of a certain background level of exertion and vigilance. In ordinary life, we are so used to exercising this level of vigilance that we do not even realize we are doing so.

With this basic picture in mind, we are now in a place to say what sanity is. Sanity is proportional to the relative strength, the relative *dominance*, of the healthy brain over the sick brain. As that proportion diminishes, as the sick brain gains ascendancy over the healthy brain, the person begins to take the pathological reports of the sick brain as veridical. The person begins to tune in to the reports that emanate from the sick brain rather than those from the healthy brain.

There are many ways that this transition, the ascendency of the sick brain, can take place. One way it happens is that the organic disease begins to spread from the sick brain to the healthy brain and gradually compromises its function. The core obstacle to the spread of disease is the corpus callosum. Thus, the callosum has two functions, not just one. It does not *merely* serve as a mechanical support for the two hemispheres, as a dense packing material. It is also a *buffer*: it stems the flow of madness from one brain to the other:

> I think myself then justified, from all I have seen, read, and manipulated, in describing the corpus callosum as merely a mechanical bond of union between the two brains, and that it does actually oppose an obstacle to the transmission of disease from one to the other, and is consequently a wall of separation rather than a bond of union.[9]

The corpus callosum quarantines the sick brain from the healthy brain—at least until the point at which the organic disease gains enough strength to breach that boundary. Wigan sometimes prefers to speak of the *sane* brain and the *insane* brain rather than the sick brain and the healthy brain: "It is the control by the sane intellectual faculties of the insane faculties; the tyranny of the sound ratiocinator over the unsound ratiocinator, that is the present object of consideration."[10]

The sick brain is the insane brain or, equivalently, the insane mind. Even when one brain is mad, it is still possible for the "person as a whole" to be sane. This is because her reasoning brain exerts vigilance, control, and dominion over the mad one. Imagine two sisters who live in a large house by themselves; one is mad, one is sane. As long as the sane one does not allow the mad one to get out of hand, as long as the sane sister enforces the proper protocols and boundaries, then we can say that the household as a whole is "functioning well." It is the same with a person. As long as the sane brain remains vigilant over the mad brain, the person as a whole is "sane," is "functioning well." As we noted, for Wigan, the whole purpose of having two brains is to ensure this possibility.

Given Wigan's identification of the sick brain with the mad brain, and the healthy brain with the reasoning brain, we can identify something like a *fundamental formula* of madness in his text. This is the root thesis, or rule, that generates his basic ontology, his system of classification, and even his method of treatment:

Madness is always a relationship between reason and madness.

The formulation has an air of paradox; it contains an apparent circularity. But this appearance of circularity is unavoidable. Madness—the kind of madness that results in a clinical consultation, or an involuntary psychiatric hospitalization—is always constituted by a kind of dialectic, an engagement or interaction between the reasoning mind and the mad mind. A looser way of putting the formula, perhaps in a more familiar but less precise way, is to say that *madness always contains reason inside of it*. We are reminded once more of Lord Polonius' quip. Madness would not be madness were it not for the fact that there is a germ of reasonableness within it. The former claim, however, *madness is always a relationship between madness and reason*, is not equivalent to the latter: *madness always contains reason inside of it*. This rather boring claim is surely held in one form or another by Kant, Locke, and Haslam. What is truly original in Wigan's thought is the further thesis that madness consists in a certain tension between reason and madness. Hence the unavoidable circularity.

Incidentally, one way to prove this latter proposition—that *madness always contains reason inside of it*—is to imagine what would happen if the organic pathology that is contained or quarantined in the sick brain, and that *makes it mad*, were to gain enough strength to sweep across the corpus callosum and completely infect both hemispheres. One might be tempted to think that such a person would be completely mad, "totally bonkers." But that would be a mistake. According to Wigan, such a condition would no longer be madness, but *idiocy*. And the distinction between madness and idiocy had been understood clearly at least since the time of Locke's celebrated formula that the madman reasons correctly from false premises, and the idiot reasons "scarce at all." Hence the paradox: madness requires a kind of inner tension between reason and madness. Madness cannot replace reason entirely without converting, immediately, into something quite alien to it.

III.

Wigan's fundamental formula—*madness is always a relationship between reason and madness*—gives us the basic ontology of madness. The formula has a corollary, which gives us his fundamental principle of classification: *the forms of madness are but different ways that reason relates to madness.* In other words, if you want to know how many kinds of madness there are, if you want a principle for "individuating" the kinds of madness, you simply ask yourself this question: what are the ways that reason positions itself in relation to madness? What are the basic attitudes that reason takes up in relation to madness? Does reason love madness? Does it detest madness? Is it indifferent to madness? There are as many forms of madness as there are stances that reason assumes with respect to madness.

We can find, in Wigan's text, three fundamental forms of madness, corresponding to three basic ways that reason can relate to madness, the three stances that reason takes in relation to madness. Recall that anytime a person is mad, it is because the sick brain has gained some power or ascendency over the healthy brain. There would be no madness at all if the sick brain had not usurped a certain amount of power. And there are at least three different ways that the reasoning brain can position itself in relation to the newly empowered mad brain. Reason can *rebuke* madness; reason can be *seduced* by madness; reason can *acquiesce* to madness. I will talk at length only about the first two forms of madness. The third, *reason acquiescing to madness*, is so perverse, so hideous, as to be nearly unspeakable; I mention it here only for the sake of completeness.

Finally, Wigan's fundamental formula gives us not only a basic ontology of madness, and a basic principle of classification, but the key to treating the mad, if and when treatment is still possible. The treatment of madness consists merely in this: the reason that lives inside of the physician *fortifies* the remnant of reason that survives inside of the patient. The physician never addresses herself to the mad brain. This is a ludicrous waste of time. The physician addresses herself to the reason inside of madness in order to strengthen and encourage it in its struggle against the mad brain, *assuming that the reasoning brain is still reasonable enough to continue its struggle.*

Reason rebuking madness. In the first form of madness, reason rebukes madness. Reason understands that the voices it is hearing, or the visions it is experiencing, or the compulsive thoughts or desires it is having, are the byproducts of a diseased organ. Though the diseased organ is getting stronger, the reasoning brain is still fighting; it is actively trying to control, direct, and contain the mad brain. Reason has not entirely relinquished its office; this form of madness is the most noble and valorous there is. This form is characterized by reason perpetually rebuking madness, remonstrating with madness, correcting madness. Imagine a mad person standing on the street screaming, cursing, yelling, apparently *to nobody in particular*. What is actually happening is that he is reprimanding his own madness; his reason is trying to silence the voice of madness, which, nonetheless, will not shut up. In a sense, he is "perfectly sane," and he is screaming at the mad mind to make it halt its mad productions. He is the apex of sanity dealing with a situation that is nearly beyond his control.

To illustrate this form of madness, Wigan produces a text that he claims captures, more or less verbatim, the ravings of a madman he listened to. The text reads like a Burroughsian cut-up, like avant-garde literature:

1. You have heard of the insults to which I have been subjected—made to lie down at the word of command—I am a gentleman— 2. The sun is my father—don't you see the rays of light from my head?— 3. They bound me hand and foot to the bed—a gentleman, sir, through ten generations—Canaille—but I'll be revenged— 4. I will rain down pestilence—dry up all their wells—give them *coup de soleil*—hats are nothing—burn them up—fry their brains— 5. Wretches! to treat me in this manner—I was entrapped here—made to believe that I was going to an hotel in my own carriage—my

own coachman that has lived with me twenty years— 6. I'll rain down red-hot pokers on them—glorious vengeance—ha! ha! ha!—set fire to the woods—kill all the pheasants— 7. Not trusted with my own gun—and you, sir, did you not give the certificate?—No, sir, I will not shake hands—it was treachery—basest treachery— 8. I shall dine off a piece of lion—calves and lions—all the same—skins make carpets—all the books in my library began to dance—you've read the Battle of the Books—Queen Mab comes here— Shakspere is wrong— 9. She is a great fat woman—they call her house-keeper—she is the only one who treats me with respect—my own little boy pretended to be frightened, and ran away from me—she never sits down in my presence— 10. Shower-bath of oil of vitriol—send them to hell—the devil told me to kill myself—he's at my ear now—begone Satan—I defy thee.[11]

It would be easy to fail to see reason in this passage; that is, it would be easy to mistakenly see this passage as an exemplar of total and complete madness, if such a thing were possible. But closer inspection reveals the unmistakable signature of reason. If you only read the odd-numbered passages, you will see a perfectly coherent train of reasoning. This form of madness can be described as a reasoning person that is persistently interrupted by a mad person.

When we take a step back, we can see that this first form of madness, *reasoning rebuking madness*, places teleology at the very heart of madness. For here, madness itself is constituted by a valiant attempt to conquer madness. Madness is not pure brain pathology. Madness is brain pathology combined with a striving to subdue that pathology.

Reason seduced by madness. This is our second form of madness. Madness has tricked reason into thinking that madness is truth. Suppose I hear a voice, apparently from out of nowhere, speaking to me. It tells me that I am a leader, that I am made for greater things, that I will lead an entire nation, if only I submit to its authority. Now, as a thoughtful person who is not mad, there are two paths before me. If I am a scientifically minded person, a progressive, secular, person, I will see in this voice nothing but the byproduct of a diseased organ and I will seek assistance. But if I am religiously inclined in even the least degree, I will waver: is this a brain disease? Or is it really the voice of God himself, speaking to me the way he spoke to the prophets? Am I like Moses before the burning bush? I find myself wavering between these two choices, but *I must choose*. If reason ultimately decides that the voice is in fact the voice of God and not the byproduct of a diseased brain, then he has given into madness. He has renounced the struggle against madness. He has been seduced and deceived by madness.

The mere fact that madness can trick reason by feigning deity, and the fact that only the religiously inclined person can be so tricked, leads us straight to the problem of religion. For clearly, my religious education, the teachings of the church that I belong to, will play a role in shaping my susceptibility to madness. And that raises the awful prospect that *certain theological systems can be more or less conducive to madness.* They can be more or less prone to make us crazy. There is no question, for Wigan, of selecting one's theology on this basis. He never says, "You must choose a theology that is the least pathogenic of all the theologies." Rather, the question is one of understanding that the theology I accept, the teachings of the church I attend, can be more or less pathogenic, and that I must, correspondingly, be more or less vigilant to the possibility that madness will use the guise of theology to subvert my reason.

Wigan, as a Protestant, addresses himself to fellow Protestants. As Protestants, he reminds them, we believe, intellectually, that the Reformation was an advance for intelligent theism; we believe that the rites and practices of the Catholics are superstitious, if not flatly idolatrous; we believe that the Holy Spirit has given each of us the capacity to read and interpret scripture as long as we do so with the humility and care that it deserves, and are supported by a community that is committed to the same. *But at least in one respect*, Protestantism has done us a massive disservice, because it encourages *all* Christians to study the Bible, to contemplate theology, to swell up their minds with mysteries that are almost impossible for anyone not formally trained in the clergy to grasp, such as the mystery of the Trinity or kenosis. The problem arises in particularly acute form if the Bible or difficult works of theology make their way into the hands of mentally vulnerable people, and particularly vulnerable young women; this will dispose them to the second form of madness, to the belief that God is somehow speaking to them; it will dispose them to being seduced by madness: "If he be in the slightest degree superstitious he considers the sensations and emotions of one of the brains as external to himself, and therefore the suggestions of a spirit, evil or good, as the case may be."[12]

How, then, to be a person of faith, if faith inclines one to madness? Should we reject religion entirely? The key is not to reject religion, not to reject God, but to *keep God at a remove*. We must keep God far away. Keeping God far away means to keep my relationship with God mediated by God's clergy, and, perhaps, if I am mentally competent, by God's word. This is what Wigan the Protestant respects about Catholics: *they understand mediation*. They do not encourage everyone to study the Bible and deep theology. They create ceremonies that stir a religious sentiment but in a tempered way. They replace the living voice of God with ritual, with incense, with incantation.

Wigan tells us that the study of scripture and of theology can make one mad. But how, precisely, does it do so? What exactly is its pathogenic feature? Wigan never says directly, but we can offer a conjecture: mentally vulnerable people, and in particular, young women, upon reading these biblical stories, would naturally wonder whether their lot in life is much greater than it actually is. Am I myself a Mary or an Elizabeth? Am I a Moses or an Elijah? Would God speak to me in the way that he spoke to his loved ones and his prophets of the past? Ultimately, the problem is not that young women will drive themselves mad by contemplating the mystery of the Trinity. Rather, they will drive themselves mad by emulating the founders of the faith. *Here I am*, said Samuel, three times. *Speak, for your servant is listening.*

In short, the diseased mind has the power to imitate God. *It is not merely misfiring*. It, too, is a thinking brain. It is a mind in its own right, and a duplicitous one. It seeks to deceive reason; it tries to trick reason into choosing madness, into relinquishing its right *to be* reason. Madness tries to tempt reason to deviate from what it is.

This structural feature of madness—that reason is always "on the inside" of madness—is the basis for sound treatment. This is particularly true of the second form of madness. This form of madness begins when *reason wavers*. At the outset and origin of this form of madness, reason wavers between the belief that these voices or feelings are the product of a diseased brain, and the belief that God himself speaks to me. The road to treatment, if treatment is still possible, is this: the reason inside the physician *fortifies* the reason of the patient. The doctor fortifies reason with reason. He defends reason with reason. He gives the patient reason to think that her visitations are not, in fact, from God, but from a diseased organ that is trying to trick her, that is trying to emulate something other than what it is, that is trying to ape God:

> There is, however, one form, or, I should rather say, one degree or stage of insanity, where argument avails something; and that is where the healthy brain is just beginning to waver in its convictions as to the reality of the delusions of its brother; where there is yet self-command when a motive is presented, and the sound organ requires to have its convictions confirmed

by something external; where, for example, the state of conscious delusion is slowly passing into that of unconscious delusion, or the disorder of one brain beginning to pass into the other; a strong appeal made to the reasoning powers of the sound brain, will, in such cases, sometimes induce and enable it to resume the reins it was just laying down in despair.[13]

The crucial point is that one never uses reason to fight madness directly. One fortifies reason in its struggle with itself.

IV.

But this mode of treatment raises a new problem, and once we understand the solution to this problem, we will understand the function of the dual brain.

Wigan's method of treatment is to fortify reason in its fight with itself. When reason wavers, when reason asks itself, *is the voice that I am hearing the voice of God, or of a diseased brain?*, it is struggling with itself. It is ambivalent between these two propositions, and it wavers between them. But if reason can waver, if it can struggle with itself, then reason itself can be divided. *One and the same mind, that is, one and the same "brain" (i.e., hemisphere), can be opposed to itself.* It is not necessary to have two brains in order to be opposed to oneself. But to acknowledge this, to admit that one and the same mind can be dual, is to abandon a core premise of Wigan's book while opening up a new window onto madness.

Recall that Wigan himself tells us, early in the book, that all of the diverse phenomena that fall under the rubric of *being-at-odds-with-oneself*, such as ambivalence, akrasia, and madness, always result from having two brains, each with its own mind. Ambivalence happens when one mind believes P, and the other mind believes not-P. So, while the individual human being is not "unified," *each brain is unified*, that is, each brain, taken alone, harbors no division within itself: "It is a diseased action of a whole brain, which cannot at the same moment have this propensity and a wish to prevent it; cannot be in two opposite and antagonist states at the same moment."[14] Each "brain" is univocal; it is in agreement with itself; it is in concord with itself. The possibility of discord within a single mind should be, within his theory, impossible. And yet, near the end of the book, he tells us that madness, or at least one form of madness, begins when the reasoning brain, the healthy brain, *wavers*, i.e., is divided against itself in its relation to the promptings of the sick brain. But to admit that a single mind can be divided against itself is to imply that a certain kind of duality, a structural duality, lives *"on the inside" of reason*. In other words, duality is not an accident of having two brains. It is a fundamental structure of reason. It is an a priori of reason. Put differently, *the mind is not one, ever*. Even if a person were born with only a single brain, that is, a single hemisphere, that person would, at least occasionally, be at odds with herself, and to that extent, dual.

With this realization, we reach our final summit. Wigan comes to acknowledge, in a roundabout and grudging way, that all cognition is necessarily dual. It has an inherent "two-ness." It has the structure of being-at-odds-with-oneself. One consequence of this is that reason and madness are both spawned from the same seed. They are two trees rooted in the same soil. Both reason and madness can only emerge from this intrinsic duality.

We are now, at long last, able to return to the *function* of this dual structure. Why do we have two brains? At the outset of the book, he presents the thesis that the duality of the brain serves the function of redundancy. It ensures that there is a backup brain in case one of them fails. But that picture has the false implication that a person born with only a single hemisphere would do "just as well," *cognitively speaking*, as a person born with both, so long as it was never impaired or sick. If the only function of the dual brain is redundancy, a person with a single hemisphere would not be at a loss, cognitively speaking, relative to a person with two hemispheres. By the same token, if you are born with only a single kidney, but your kidney functions optimally for your entire life, you would not be at a loss relative to a person with both.

By the end of the book, however, we discover that having two brains is not just a matter of functional redundancy. It is essential to mind itself. It marks the difference between mind and what he calls instinct. This is a startling admission. At first, each brain had one mind, and each mind worked perfectly well all alone, but we needed a backup, and that backup makes us prone to madness. But now he goes on to admit that this duality has a positive role to play in reasoning itself. In fact, a person with two brains can reason better than a person with only one. What is the cognitive benefit of having two brains? Having two brains allows me to take two different stances on one and the same topic, to compare, to have a dialogue with myself, and thereby reach a more considered conclusion. Philosophers know this all too well; we play devil's advocate with ourselves in order to ascend to a more exalted truth. This, in fact, is the crucial difference between human and beast:

> May it not be, that the distinction between mind and instinct consists in
> the parity or disparity of the two cerebra. If exactly equal and alike in every
> respect, does not the animal necessarily act uniformly from instinct? that
> is, from uniform unvarying impulse. Whereas, if unequal in power, and
> slightly different in function, the ideas of one brain may be weighed against
> the ideas of the other, there would then be comparison and judgment, and
> a progressional advancement in knowledge,—and this, whatever be the
> shape of the organs of intellect.[15]

Wigan goes so far as to entertain the proposition that our very ability to empathize with another person stems from the duality of the brain. Consider
a game of chess. With one brain I plot out my winning strategies; with the
other brain I adopt the stance of my opponent and anticipate what he will
do next.[16] While Wigan does not ultimately accept the thesis that empathy
requires this duality of the brain, he does accept the more general proposition that there are distinctive cognitive benefits associated with having two
brains.

Finally, in light of the idea that reason requires this two-ness, Wigan insists
that the most stable kind of person is not one whose hemispheres are equal
to each other in power. Equality of hemispheres would lead to anarchy, to
psychic chaos. There must be a natural hierarchy between them; there must
be a superior one and an inferior one; there must be a relation of dominance
and submission. Wigan thinks the left brain is typically the dominant one,
because it controls the right side of the body and we do so many things with
the right side of our bodies. So, while reason and madness both require this
two-ness, the sane person is one who has established a settled hierarchy, a
dominance, of one brain over the other.[17] Madness shows itself to the extent
that this natural inequality is flattened or reversed; it is the absence of the
right kind of hierarchy.

V.

As I alluded to above, there is a third source of madness, a third form of
madness, *reason acquiescing to madness*. What makes this different from
the second kind of madness, *reason seduced by madness*, is that, in the latter,
reason is tricked by madness into abdicating its proper authority. Madness

feigns the voice of God; reason, wavering and at war with itself, ultimately takes a leap of faith and freely decides to choose madness. But this third form of madness is quite different. Here, *reason understands madness to be madness but chooses madness anyway.* Wigan says very little about this kind of madness; he only raises the possibility of this twice, and in a passing way. It is almost as if this third form is so repugnant to nature it can scarcely be thought, much less spoken of.

He gives us two examples of this. First, he tells us of a man who lost all of his wealth through speculation and went completely mad. In his madness, he formed the delusional idea that he was still rich. He spent his time writing imaginary paychecks to his friends and to charities that he loved. As far as madness goes, there would seem to be nothing atypical here, but for the fact that *the reasoning part of his brain understood perfectly well that he was destitute but chose madness anyway*:

> He now fancied himself possessed of immense wealth, and gave without stint his imaginary riches. He has ever since been under gentle restraint, and leads a life not merely of happiness, but of bliss; converses rationally, reads the newspapers, where every tale of distress attracts his notice, and being furnished with an abundant supply of blank checks, he fills up one of them with a munificent sum, sends it off to the sufferer, and sits down to his dinner with a happy conviction that he has earned the right to a little indulgence in the pleasures of the table; and yet, on a serious conversation with one of his old friends, *he is quite conscious of his real position, but the conviction is so exquisitely painful that he will not let himself believe it.*[18]

In this case, madness is a more or less deliberate diversion or escape from the unremitting tragedy of everyday life.

He raises this prospect only once more. He describes a case of a woman, "a pauper lunatic who believed herself to be Mary Queen of Scots."[19] In giving his explanation, he tells us, "The more sordid and sad were her occupation and employment, the more (probably) was she tempted to indulge in the pleasures of imagination—as players take refuge from the wretched cares of their precarious life in a sort of half-belief that they are the characters they represent."[20]

This raises a significantly new picture of madness, one in which madness is a strategy for escaping from a painful reality. It is a "coping mechanism." What is distinctive and troubling about this new vision of madness, which we will shortly revisit in more detail, is that reason prefers madness to itself, while recognizing it *to be* madness, because madness offers an advantage that reason simply cannot give it. Madness is a tool for getting a job done.

8

Delusion as Castle and Refuge

Johann Christian August Heinroth, in his 1818 textbook of psychiatry, *Lehrbuch der Störungen des Seelenlebens*, raises the possibility that some forms of madness are coping mechanisms, that is, attempts by the mind to flee a painful reality. In Heinroth's view, a wise healer is one that *allows nature to run its course*; in the best of circumstances, the patient will eventually return to normalcy on her own (Section I). In the early twentieth century, psychoanalytic psychologists developed Heinroth's ameliorative model in a systematic way and applied it to the treatment of schizophrenia. Frieda Fromm-Reichmann—coiner of the phrase "schizophrenogenic mother"— taught that schizophrenia was not *just* a flight from a painful reality, but more specifically, an escape from repeated experiences of early childhood rejection. The key to therapy, then, was to lavish unconditional acceptance and approval upon the patient (Section II). For Fromm-Reichmann's colleague Harry Stack Sullivan, schizophrenia is also a sort of purposeful inner journey, but one with a very specific goal: to retrieve and reincorporate life experiences that had been dissociated (Section III). This conception of schizophrenia as a strategic retreat from painful experiences raises a question that cannot be entirely answered in this chapter: if schizophrenia is designed to be a healing journey, *why does it so often fail*? Why does it ever become chronic? We will return to this problem when we approach Laing.

I.

In his *Memoirs of My Nervous Illness*, Judge Daniel Schreber describes how the nerves, the "finest filaments" or "rays" that constitute God's mind, are attracted to the nerves or "rays" that constitute the human mind. The problem is that for God to make nerve-contact with a human—a living, embodied being—actually jeopardizes his very existence, for it is possible that once contact is made, God would not be able to extricate himself. At the time that Schreber penned the manuscript in 1900, he was unable to

Madness. Justin Garson, Oxford University Press. © Oxford University Press 2022.
DOI: 10.1093/oso/9780197613832.003.0009

say exactly *why* God could not extricate himself. But a footnote added in November 1902 illuminates the situation: God's inability to extricate himself would not be due exclusively to a quasi-mechanical obstruction, for example, tangled filaments, but also, in part, to the intensity of his own desire for humanity; it is not that he cannot, *but that he does not want to*, extricate himself:

> This phenomenon will perhaps be somewhat more comprehensible and brought nearer human understanding if one remembers that the rays are *living beings* and therefore the power of attraction is not merely a mechanically acting force, but something like a *psychological motive power*: the rays, too, find that "attractive" which is of interest to them. The relationship therefore appears to be similar to that expressed by Goethe in his "Fisherman": "partly she dragged him down; partly he sank."[1]

This situation, this possibility of entanglement, also gives us an image of a new way of thinking about madness, not as a disease that overpowers one, that drags one in against one's will as a riptide carries one out to sea, or as an involuntary eruption of infantile or collective contents; we must instead confront the possibility that one might *prefer* madness to reason, that one might choose madness over reason while at the same time *recognizing madness for what it is*. One is neither tricked nor duped by madness; one freely acquiesces to it. This is the possibility that Wigan raises yet recoils from all too quickly. The bare possibility that a reasonable person might *choose* madness over reason appears to be absent before the nineteenth century. True, Shakespeare's Constance, in *King John*, exclaims that madness is preferable to reason. In grief over her murdered son, she says:

> I am not mad: I would to heaven I were!
> For then, 'tis like I should forget myself:
> O, if I could, what grief should I forget!

Shakespeare gives voice here to the perfectly coherent view that madness could be preferable to reason, just as death could be preferable to life. Unlike death, however, one cannot willfully step into madness. To the extent that one "chooses" to be mad, it is always on the order of a ruse, as in Hamlet. And while Burton recognizes the pleasure of building his "castles in the air," he never once suggests or intimates that madness could happen because one deliberately chooses to "step into" the very castle that he or she has created.[2]

At the turn of the century, however, this possibility is broached by Johann Christian August Heinroth, the first person to hold a chair in psychiatry in Europe. Heinroth was also the first to use the term *Psychiatrie* in a book, his *Lehrbuch der Störungen des Seelenlebens* of 1818.[3] (His mentor, Johann Christian Reil, had coined the term ten years earlier.) For Heinroth, as for Burton, madness is a form of sin and a result of sin. In a normal individual, consciousness evolves through three "grades": *world-consciousness* (*Weltbewusstseyn*), *self-consciousness* (*Selbstbewusstseyn*), and the highest level of consciousness, *reasoning consciousness* (*Vernunftbewusstseyn*), which he also calls a mode of *non-self-existence*, because it carries within it "the sense of the infinite, the unlimited, the eternal."[4] Few attain to the final stage in the evolution of consciousness, yet all can move in that direction. Health is a question of moving in the right direction. And what draws us onward and upward is the voice of conscience, God's implanted radio device. When we "tune in" to the voice of conscience, we can follow its prompting and start to enjoy a "wonderful, strange purity and clarity, which persists for only as long as we do not think about the world and about Self."[5] With patience and fortitude, obeying the promptings of conscience will carry us to a state of ultimate psychological health, in which

> the whole man is filled with and pervaded by a joyful, active vitality. The feeling of this harmonious, serene vitality, the flight of which cannot be compared with any feeling of pleasure on a lower plane, is the true state of human health.[6]

It is also possible, however, for a person to "stall out" at an earlier level or even regress. And this stalling or regression is not an involuntary affliction, like a developmental disorder. It is a choice; it is an expression of one's evil will. The prodigal son is not quite ready to turn; the life of sensual indulgence has not yet betrayed him, has not yet shown its hollowness. And this condition, this willful stalling out or regression, is a kind of madness; it is, in fact, the entirety of madness:

> The man who is fettered by passion deceives himself about external objects and about himself. This illusion, and the consequent error, is called madness [*Wahn*]. Madness is a disease of reason and not of the soul, but it originates from the passion within the soul. . . . Man cannot be freed from madness until he is freed from passion. In madness the spirit is fettered

and man, just as in passion (both being indissolubly linked), is unfree and unhappy.[7]

So all madness, in the broadest sense of the term, is dysteleology; it is a willful disruption of the purposeful and goal-directed movement of the soul toward reasoning consciousness.

Strangely enough, after Heinroth sets forth this moral vision of madness, his textbook reads as a fairly conventional, even Kantian, medical treatise. There are three components of the soul (*Seelenwesens*): mind, spirit, and will (*Gemüth, Geist, Wille*). There are three ways that any faculty of the soul can be disturbed: through "exaltation, depression, or a mixture of the two." Hence, we have *nine basic categories of madness*, since there are three ways that each of the three components of the soul can err.[8] In addition to these *simple* forms of madness, there are hundreds of *compound* forms; there is a complex calculus of the forms of madness. In spite of his theological starting point, it would be understandable if one reads Heinroth's textbook as an expression of madness-as-dysfunction—and in general, it is.

There is one particular form of madness, however, that breaks that mold and raises a prospect that he had not previously raised. (Schmorak, in his English translation, translates this form of madness as "insanity with dementia and rage" [*Wahnsinn mit Verrücktheit und Tollheit*].) This form of madness is a way of retreating or withdrawing from a lifetime of suffering, tragedy, ridicule, and disdain, and entering into a kind of dream world. It is a compensatory reaction, a coping device, a strategy for healing:

> In a life which has been depraved by neglected education, circumstances, and guilts, in which excited sensuality, perverted concepts, and prejudices are prevalent, in which the intellect has been totally neglected or has been strained by brooding or spurious speculations, in which unrestrained arbitrariness rules, in which many embarrassments, restrictions, inhibitions, and dangers have been the lot of the patient, in such a life a moment may come when the measure is full and runs over, and the imagination becomes strained to the limit in trying apparently to effect a full compensation and to transform the evil fate as though by the touch of a magic wand.[9]

This is not to say that *Wahnsinn mit Verrücktheit und Tollheit* is a *conscious* strategy; I am not consciously willing this to happen. But whether conscious or unconscious, it is a strategy that the mind deploys to escape the horror of reality. In fact, in a medical textbook that runs to almost 800 pages, that is the only disease that he explicitly recognizes as having a teleological dimension, as governed by a purpose rather than a failure of purpose.

But what is the fate of a person who has chosen madness in this way? In the best of cases, this journey into a dream world has a cathartic function; it is not a terminus, but a way station toward a higher sanity. In these cases,

> reality may reassume its rights, the intellect may recover, and the patient is again fully conscious and is able and willing to master his own morbid will-fulness. In such a case, the disease acts as a storm which clears the atmos-phere, a wholesome desire of nature to cure a perversion through another perversion [*ein heilsames Bestreben der gesunden Natur, Verkehrtes durch Verkehrtes zu heilen*].[10]

In this case, the episode of madness actually propels the soul along in its movement toward reasoning consciousness; it helps her to master her will-fulness; she can now see the world with fresh eyes. She has come to her senses and is ready to take on the challenges that life constantly throws in her path; she is "up to the task." So, we can actually see in this form of madness two levels of teleology in operation. On the one hand, the disease is initiated by a certain strategy, namely, *I just want to escape from this terrible reality*. But then it turns out that the entire episode—from immersing oneself in the dream world, to returning renewed and refreshed to reality—has been orchestrated by the wisdom of nature. The entire process is *nature's* way of curing the pa-tient. The patient may have initiated this form of madness through her re-jection of reality, but nature has scaffolded and carried her to its felicitous conclusion. Yes, it is a perversion, as all forms of madness are, but nature has allowed this perversion to take place; nature has, in its benevolence, created a space for a violation of the natural order. Imagination *should be* subordinated to judgment, as Burton puts it; that is the order of nature. But in this partic-ular case, nature has made an exception; nature has allowed the imagination to gain the upper hand as an expedient to healing.[11]

This procedure, nature's cure, is not without risk. The painful truth is that this form of madness does not always culminate favorably. Sometimes the cure fails. It occasionally happens that one enters the dream state, and never leaves it. Those cases gradually pass over into a dementia from which it is im-possible to recover. It also happens, in some rare cases, that the dream state culminates not in dementia but in a more or less permanent paranoid state, a rigid set of delusions. Finally, it is possible that one does return, but with one's willpower and emotions shattered. Later psychiatrists will revisit this theme repeatedly: if madness is designed to be a kind of healing journey, *why does it so often fail?* What goes wrong in the inner journey such that madness becomes chronic? As Gregory Bateson writes in his introduction to *Perceval's Narrative*:

> It would appear that once precipitated into psychosis the patient has a course to run. He is, as it were, embarked upon a voyage of discovery which is only completed by his return to the normal world. . . . In terms of this picture, spontaneous remission is no problem. This is only the final and natural outcome of the total process. What needs to be explained is the failure of many who embark upon this voyage to return from it.[12]

With *Wahnsinn mit Verrücktheit und Tollheit*, a new form of madness has been created, and with it, a new role for the healer: the physician's job, in the first place, is not to subvert or undermine nature's work. When the physician sees the patient detaching from reality, the physician's goal is not to wrench him back out of it with purging, bleeding, vomiting. Nor is the physician *simply* to allow nature to take its course. Rather, the physician must become a shepherd and guide. This role is only articulated in a very explicit way with Laing. But it is there, in Heinroth, as an implicit possibility.

II.

At this point we are forced to depart from the roughly chronological progression this book has followed so far, and jump ahead by a century, to the work of Frieda Fromm-Reichmann and Harry Stack Sullivan, two psychiatrists who, in the first half of the twentieth century, attempted to bring Freudian insights toward the healing of psychotic patients. The justification for engaging with their work here, rather than in a separate chapter, is the striking resemblance that their thinking bears to Heinroth's: for Fromm-Reichmann and Sullivan, schizophrenia is a coping mechanism, an escape from the world, but also, and at the same time, a mode of engagement with the world.

By the first half of the twentieth century, this vision of madness as a kind of refuge and castle returns in Oedipalized form. Madness—or schizophrenia, as it is now called—is not a result of the generalized misery of existence, but a retreat from the trauma of constant early rejection. It is a response to an inhospitable family. For Fromm-Reichmann, coiner of the term "schizophrenogenic mother," this is axiomatic. The following statement can be taken as the theoretical foundation of her entire approach:

> The schizophrenic is painfully distrustful and resentful of other people, due to the severe early warp and rejection he encountered in important people of his infancy and childhood, as a rule, mainly in a schizophrenogenic mother. . . . The schizophrenic's partial emotional regression and his withdrawal from the outside world into an autistic private world with its specific thought processes and modes of feeling and expression is motivated by his fear of repetitional rejection, his distrust of others, and equally so by his own retaliative hostility, which he abhors, as well as the deep anxiety promoted by this hatred.[13]

We should be quick to add that—as emphasized by her biographer[14]—this passage represents the only place in her entire corpus in which she uses the phrase "schizophrenogenic mother." The idea of the schizophrenogenic mother, per se, was not the centerpiece of her philosophy of madness and its healing, and it is unfortunate that her "place" in the history of psychiatry has been so closely aligned with this label. In other parts of her corpus, she clarifies that the instigator for the "schizophrenic withdrawal" is not always the mother; rather, it differs from culture to culture. In Western Europe, a deeply patriarchal culture, the instigator is more commonly the father; in America, "where women are often the leaders," it is usually the mother.[15] In some cases she is clear that the trigger for withdrawal is a maladaptive pattern of relationship between mother and father.[16] The critical process that we must understand, in order to understand Fromm-Reichmann's work and practice, is that this withdrawal is motivated by an early experience of *what the patient interprets as* an act of rejection by significant others, and *the entirety of the therapeutic process is designed to redress the effects of this early, perceived rejection.* At base, Fromm-Reichmann invites us to admire the inherent *reasonableness* of madness. Schizophrenia is a form of social retreat motivated by the inherently reasonable goal of avoiding future rejection.

A crucial feature of the relationship between therapist and patient is that *they are not living in the same world.* The therapist is living in one world, "our" world, and the patient is living in another, a "private" world, an inner world. And the therapist's job is to establish a bridge, "a bridge over which [the patient] might possibly be led from the utter loneliness of his own world to reality and human warmth."[17] She must penetrate into the patient's world without disrupting it—that is, while respecting its intrinsic stability and raison d'être—and establish a lifeline back into "our" world.

But how is such a rescue operation even conceivable? If the patient inhabits a different world than I do, we can speak figuratively about "entering into his world," but in fact, he has put himself *out of communication*: his world is not accessible from ours; in relation to ours, it is a black, sucking void. The defining feature of that world, his world, is its inaccessibility. It is not just another region in a kind of imaginary space, but a hole in that space.

It *would* be impossible to enter his world, *but for* the fact that the mad person is himself dual, is himself ambivalent, about his peculiar situation. As a human being, he still has an instinct, a drive, toward sociality. As a human, his instinct toward withdrawal can never be complete. There must be, inside of him, not only the rejected child, but an adult who says, "I miss the world; I miss connection; I miss my family; I miss my friends."

> In spite of his narcissistic retreat, every schizophrenic has some dim notion of the unreality and loneliness of his substitute delusionary world. He longs for human contact and understanding, yet is afraid to admit it to himself or to his therapist for fear of further frustration. That is why the patient may take weeks and months to test the therapist before being willing to accept him.[18]

It is for this reason that the healer must address herself to *multiple persons at once*: a frightened little child, and an adult who longs for connection, interaction. In fact, addressing oneself exclusively to the child is apt to subvert the whole process, because the patient, the adult part of the patient, is aware of and very sensitive to condescension. After all, his whole dilemma came from his being repeatedly rejected, so even though he appears to be in his own world, he is still sensitive and vigilant to the threat of rejection or ridicule. He let down his guard once and is not going to let it happen again.

We are now faced with a seeming contradiction of two different functions, two contrary functions played out simultaneously and completely. The first is withdrawal, detachment from reality, being out of communication. Yet there is also a being that exists somehow alongside this first being, who is not withdrawn but *fully and completely present*, who is monitoring his outer world in extraordinary detail:

> The schizophrenic's expectancy and tendency toward resuming interpersonal contacts were sometimes equally as strong as his original motivation for withdrawal. This seemingly paradoxical attitude can be easily understood since the schizophrenic has not resigned from interpersonal dealings freely or of his own design but is motivated by dire, defensive necessity.[19]

There is yet another reason that this apparently contradictory situation, simultaneous withdrawal and presence, is nonetheless *necessary*, that is, why they have to coexist. One withdraws from the world for the purpose of creating a kind of sanctuary, a castle. But the only way I can relax fully, the only way I can take refuge in my castle, the only way the refuge can *be* a refuge, is if I feel fully confident that I am safe, that I am not going to be invaded by an outside force. So, I need a watchman; I need someone to remain watchful and vigilant, to keep a lookout and disarm any threats that appear to be moving in my direction. Hence the necessity of two beings, yoked together; this necessity is what allows for therapy. If it were possible, if it were even imaginable, that a person could decisively and utterly withdraw from reality, then therapy would be impossible.

Because of the mad person's vigilance, therapy is not only a risk for the mad person, but for the sane person, the healer. The patient is not only prone to withdraw if he feels any condescension, but he is also extremely observant, acutely aware of any signal of insecurity or weakness coming from the healer. And as his job as watchman is to disarm potential threats, he is capable of subverting the therapeutic process by attacking, with laser-like precision, the therapist's own social vulnerabilities:

> The schizophrenic's ability to eavesdrop, as it were, on the doctor creates another special personal problem for some psychiatrists. The schizophrenic, since his childhood days, has been suspiciously aware of the fact that words are used not only to convey but also to veil actual communications. Consequently, he has learned to gather information about people in general, therefore also about the psychiatrist, from his inadvertent communications through changes in gesture, attitude and posture, inflections of voice or expressive movements. . . . Therefore the schizophrenic may sense and comment upon some of the psychotherapist's assets and, what

is more frightening, his liabilities, which had been beyond the limit of the psychiatrist's own realization prior to his contact with the schizophrenic patient.[20]

Hence, what is called for is a simple connection between two adult, reasonable human beings about the situation. The therapist must be truthful, trustworthy, candid. Nonetheless, while addressing herself to the reasonable, vigilant, outward-looking adult, the therapist must also and at the same time address herself to the rejected child. In order to have an audience, as it were, with the scared child, one must demonstrate one's trustworthiness. Thus, to the extent that it is possible within the limits of the therapist's job, one must never refuse or rebut or reject the patient. One must have extended hours. If the patient cannot make it to therapy, one must not reprimand them. Where the mother was a perpetual *no*, you must be a perpetual *yes*: treatment is continued with "as much acceptance, permissiveness, and as little rejection as could possibly be administered."[21] The problem, of course, is that if you address yourself too much to the scared child through your *yes*, then the chronologically aged adult will see your ruse and feel that he is being condescended to, that you are treating him as if he were a child. So, it is an extraordinarily delicate act—she reminds us of Federn's advice that "when we treat a schizophrenic, we treat several children of different ages"[22]—but not by that means impossible.

The idea that a person with schizophrenia is in a "different world" was the basis of Freud's pessimistic verdict that in the case of schizophrenia, psychotherapy, in its then-current form, could not help.[23] This is not to say that Freud held the same theory as Fromm-Reichmann; he did not. For Freud, the schizophrenic has, in a sense, withdrawn into himself. But he has not withdrawn into himself because he is afraid of some external event or person— afraid of the mother's rejection, afraid of tragedy, afraid of loss. Rather, he is afraid of *himself*; he is afraid of his own longings. Suppose a man feels an intense sexual longing for another man, but he cannot consciously confront the depth of his desire. His libido is attached to another man but he cannot make himself aware of that. He detaches his libido from his love interest. But where, then, does that extra libido go? For it must go somewhere; libido is a fixed quantity. Effectively, the mad person turns his libido *inward*, onto himself; this is narcissism. And when libido is so withdrawn from the external world and directed toward a person's own self, there ensues a sense of de-realization, a loss of reality: this is the birth of Schreber's "fleetingly im-provised" men. So, for Freud, the patient is no longer attached to the external world in the way that would be required for psychotherapy. The patient has fled into his castle and burned his bridges behind him.

III.

For Harry Stack Sullivan, too, Fromm-Reichmann's friend and colleague, *madness is always thought of teleologically.* The starting point for the psychia-trist is not *dysfunction*; it is *strategy*. It is astonishing that Sullivan was able to reflect on the course of psychiatry in the previous century and formulate this point explicitly:

> The field of study being *mind*, any major tendency of scientific thought to pass over to philosophy must reside in the nature of mind itself. What seems to be the principal feature of mind which stimulates this generally unwitting transition? Now that the enthusiasm for quasi-scientific ob-scurantism is failing with the collapse of materialism in the joint fields of physics and biology, it seems reasonably safe to express, as an answer to the question, the obvious fact of *teleology* as a characteristic of things mental. . . . In attacking the problem of understanding and treatment of schizophrenia, the element of motivation seems logically fundamental to all others. As Dr. William A. White has said, "We must understand what the patient is trying to do."[24]

Moreover, this teleological orientation is *not merely opposed* to "sterile brain physiology," but also to "psychologization." Although Sullivan does not elab-orate, at this point, what this "psychologization" amounts to, we can take it for granted that when he takes the *teleological* as the starting point of psychi-atry, he is not, or not merely, opposing mental and physical, that is, the *psy-chological* approach and the *biological* approach. He is urging that we accept a very different starting point, a different principle of division: teleology and its failure.

In articulating his therapeutic orientation, he tells us that the "point of de-parture" for his own research was a passage written by Adolf Meyer.[25] The "most significant text" is the following:

It is not by any means excluded that the capacity to go into a catatonic reaction may be looked upon as a positive asset, i.e., not as a product of the "disease," but as a defense mechanism indicative of the constitutional makeup. With a more adequate study of the mechanism it may be easily be found that the forces at work in the reaction as such might be turned to the use of the therapy, as undoubtedly nature "intended" to do.[26]

What, then, is the purpose of schizophrenia? What did nature intend by it? Like Fromm-Reichmann, Sullivan sees it as a strategic retreat into self, but unlike Fromm-Reichmann, it is *not merely* a matter of fleeing from early experiences of rejection, of "the direct, carefully focused, and terribly warping hostility to which a good many preschizophrenic children are exposed."[27] Rather, a person with schizophrenia is regressing to an infantile state *for the purpose of recovering something that she lost*, and something that she desperately needs in order to function *as* a person. The withdrawal into the self is not merely a question of fleeing into a refuge or a castle; it is a recovery operation. But what has been lost? Sullivan tells us that

> by 1925 I had convinced myself of the inadequacy of *any* extant formulation of the schizophrenic states, and offered a preliminary statement of the conservative as contrasted with the destructive aspects of these conditions: "the conservative aspects . . . are to be identified as *attempts by regression* to genetically older thought processes . . . *successfully to reintegrate masses of life experience* which had failed of structuralization into a functional unity and finally [had] led by that very lack of structuralization to multiple dissociations in the field of relationships of the individual not only to external reality, including the social milieu, but [also] to his personal reality."[28]

Schizophrenia is a strategy for regressing to earlier life stages for the purpose of retrieving dissociated "masses of life experience." So, what are these dissociated masses of experience? Why did they become dissociated? How to get them back?

To understand the goal of schizophrenia, we must begin by understanding the self. The *self* is a kind of "social construct"—if that term has not been entirely evacuated of meaning by now. The self is a product of social interaction. The self is "invented, evolved, tooled, and refined for the purpose of making us feel secure among our fellows, getting what we want from cooperation with them, and avoiding their more open hostilities."[29] The self is an artifact of the process of interacting with others, getting one's needs met, etc.

The basic mechanism that gives rise to the self is this: as a child, I perform certain actions and I see that they draw approval from parents or other important figures; I perform other actions and I see that they evoke disapproval. That self, the one that meets with approval, is the *good me*; that other self, the one that meets with disapproval, is the *bad me*. But there are certain things that I do, like masturbation, that meet with such intense disapproval and anxiety that I cannot see *myself* as the author of those doings at all. Those dispositions or capacities or tendencies are shuttled off into a secret realm: this is the *not me*. This is how the self is sculpted through its interactions with others all the way through adolescence.[30]

The problem is that every time I shuttle off a disposition or propensity (the "not me"), I open the door to future maladjustments. And this for two reasons. First, those propensities or dispositions can *continue to be operative* even in a suppressed and unconscious state. As outside of my conscious control, they can be troublesome. They demand to be exercised, fulfilled, and they have the power to insert themselves into my behavior in ways that can cripple my ability to function normally in society. They appear as unwelcome intrusions, both into my interactions with other people as well as into my thought life. But second, and more deeply, some of these dispositions are absolutely vital to living a fulfilling life. Without them, we cannot be, as Sullivan puts it, *fully human*. These maladjustments can represent anything from mild irrationalities to full-blown psychosis.

Sullivan gives an example of how this shuttling off of personality dispositions can create mild maladjustments in one's everyday life. A young man has a positive sexual encounter with another young man; ashamed, he quickly represses and disowns these desires; they are shuttled off into the *not me*. These desires are still operative inside of him, but now as "out of circulation," as repressed. Nonetheless, the desire is still active; it still seeks satisfaction; it still guides his action. Now, as an older man, he has a colleague who resembles his young partner, and "the resemblance would in any ordinary instance be utterly sufficient to provoke a vivid recall in [him] of the early, highly satisfactory experience."[31] Instead of desire, however, he feels nothing for him; he feels indifference; in fact, he is not just indifferent but he feels mildly irritable and annoyed toward his colleague. If someone points this out to him, and if he is honest, he might say, *I don't know what it is but there's something I just don't like about him. Just a feeling, I guess.* That represents the *self-system* striving to repress his dissociated desire by converting it into its opposite. The self-system, in a sense, *is not pathological*; it is functional; it has a goal, aim, purpose. The self-system has a task to accomplish, a job to do; as a rule, it is crucial to social stability. But now its operations have created a mild maladjustment. He cannot work effectively with his colleague because the dissociated tendency is banging at the door, and the self-system must neutralize and negate that tendency by emphasizing the opposite feeling: indifference, even irritation.

What we have described here is simply a mild maladjustment, not full-blown psychosis. But the same mechanism can explain full-blown psychosis, too. Suppose I have been forced to dissociate major motivational systems: sexuality, human warmth, human communication, "a thing as essential to being human as a major integrative tendency."[32] Suppose every time I have sought to express intimacy, warmth, or affection toward a caretaker, I have been met with disapproval, even revulsion. If I have been forced to dissociate those feelings, I am going to be severely disabled when it comes to forming normal interpersonal relationships. This disability could express itself in as sharp a form as schizophrenia. One and the same mechanism creates both mild neuroses and full-blown madness, namely, the dissociation of "integrating tendencies," that is, the disowning of certain basic forms of interaction with other human beings.

With these ingredients in place, we can finally understand the nature of the inner journey that constitutes schizophrenia. The schizophrenic journey is initiated when the self-system, under duress by a life situation, is *strained* to the point where it can no longer function effectively to keep the dissociated contents out.[33] The wall of self becomes thin, permeable. Dissociated contents begin crowding in. Many of these contents are ones we shuttled off during infancy; their bizarre or incoherent character is simply their infantile character. In schizophrenia, we glimpse the world of the infant. In this world, bodies, things, our environment are not segregated neatly: "It is not necessary nor yet possible for the child to be quite clear on the demarcation between I, myself; mama; her nipple; the dog; the inanimate Teddy-bear; and so on and so forth."[34] Me-mama-nipple-dog-teddy is one expansive "thing."

Crucially, schizophrenia is not *constituted* by this "overflow" of dissociated contents, this rupture in the self-system, for in that case, schizophrenia would be *merely* a disease, a pathology, a breakdown of function; there would be nothing of teleology in the matter. Rather, this overflow is what instigates the journey of schizophrenia, the basic goal of which is to reincorporate some of those dissociated contents. The nature of the breakdown points the way toward its own solution.

Now we are on the journey, the recovery mission. The problem is that the recovery mission is both perilous and anxiety-provoking. There is a very good reason that we dissociated those masses of life experience: they evoked disapproval and caused anxiety. Moreover, it is completely unknown to me what my life will be like when those dissociated contents are reattached to my personality. In short, "the metamorphosis is scarcely an attractive prospect."[35] And it is because of this that the journey is so perilous. It is perilous because, as we embark on it and come closer to these basic sources of anxiety, these infantile sources of anxiety, we resist; we waver; we want to give up. And there are various ways of giving up. The chief danger on the journey of schizophrenia is that we give up prematurely and get stuck in a chronic inner madness. And these ways of getting stuck, these ways of aborting the schizophrenia journey, *are themselves subtypes of schizophrenia.* The essential form of schizophrenia, the goal-directed form, the *good* form, is the catatonic type, and *when the catatonic "mission" is aborted we get either the paranoid or hebephrenic type,* depending on the particular strategy that the patient is utilizing to deal with the anxiety stirred up by the inner journey. But the paranoid and hebephrenic forms are degenerations of the catatonic form; they are, in a sense, deviant reactions to the valiant journey that constitutes the catatonic form. Such patients tend to become chronic unless they are "pushed" back into the catatonic form.

The *essential* form of madness, its truest and purest form, is catatonic schizophrenia, marked by almost total social withdrawal, involving "disturbances of motion, or skeletal activity, and by such phenomena as mutism, the refusal of food, and neglect of the toilet habits."[36] And Sullivan can see it as the essential form because it is *just as if the person is no longer there;* he has departed for a long journey, a journey to recover dissociated tendencies of the personality: he has become a hole in social space. But how does this essential form become converted into the degenerate forms, the hebephrenic or paranoid forms?

The *hebephrenic* is marked by "dilapidation of social habits, seclusiveness, and irregular episodes of incredibly purposeless destructive excitement."[37] Here we have all of the symptoms that go under the vague rubric of "thought disorder": incoherent speech, tangentiality, thought blocking, pressure of speech, etc. Suppose I am on this schizophrenic journey (the good, catatonic sort) and I decide that I will not succeed. *I will never be fully human. I will never be able to engage normally with other people. I give up.* That is the hebephrenic form of schizophrenia; it is simply despair: "If the schizophrenic despairs, he becomes hebephrenic."[38] I withdraw from others, I mutter to myself, I begin to take pleasure in the most rudimentary sorts of bodily functions: playing with my own shit, holding in my saliva until it leaks out of the corners of my mouth. It is a way of saying: *this schizophrenic journey is too hard; it's probably never going to work; I will never be a socially well-adapted person; this is about the best it's going to get for me.*

Alternatively, I can adopt the paranoid solution. This is the solution that is particularly attractive for those who are racked by guilt and shame. "The pre-schizophrenic wants to be human, and cannot find out how to do that. The pre-paranoid schizophrenic wants, in addition to that, to be blameless, to be rid of things of which he is profoundly ashamed and which he regards as part of his handicap in being human."[39] It is a way of saying: *the problems I'm having aren't my fault. It's not because of my own failures or shortcomings or weaknesses. It's their fault. They are the ones that are doing this to me. They are the ones putting me in this situation; in fact, it is probably because I am such an important person that they are bent on destroying me in this way. They are putting thoughts in to my mind; they are poisoning my food; they have cast me into exile.*

In short, Sullivan recognizes two very general categories of madness, which we will call the *good* madness and the *bad* madness. There is the *good* madness, which is positive, life-giving, therapeutic, healing. Then there is the *bad* madness, the perverse, the chronically maladjusted, and even the ultimately untreatable. And the bad madness—hebephrenic and paranoid schizophrenia—happens when the schizophrenic journey is aborted because the sufferer loses hope that he will ever recover, or because he needs to evade the guilt that his journey evokes. If we think of the schizophrenic journey as itself a goal-directed process, as having a target, then the bad forms of schizophrenia—the hebephrenic and paranoid forms—are ways of missing that target; they are ways of going awry. In that sense they represent dysteleology in relation to the teleology of the catatonic form. Nonetheless, even though the paranoid and hebephrenic forms of schizophrenia represent dysteleology, *ways that schizophrenia can go wrong*, they are nonetheless still goal-directed and purposive in their own right because they represent *solutions*. They are cop-outs, to be sure, but even cop-outs are solutions in their own right; they are easy solutions, quick fixes:

For those whose personal history permits it, the elaboration of a paranoid distortion of the past, present, and future comes as a welcome relief. . . . It is an improvement, so far as the security of the patient is concerned, not only on the catatonic state, but on the previous uncomfortable prepsychotic existence. Therein lies its most evil potentiality. It can lead to nothing conducive of personal development; quite the contrary. But it can and does give a pay-as-you-go security.[40]

As a rule, the paranoid and hebephrenic forms of schizophrenia settle into a more or less chronic state, and this is their aim: "A paranoid systematization is, therefore, markedly beneficial to the peace of mind of the person chiefly concerned, and its achievement in the course of a schizophrenic disorder is so great an improvement in security that it is seldom relinquished."[41] In these cases, however, it is possible—albeit dangerous—to *compel the paranoid patient to return to the catatonic state*:

I have, in the days when my recklessness was perhaps paving the way for a little sanity later, upset pretty elaborate paranoid systems by nothing more startling than a warm personal attitude toward the patient, combined with sundry attacks on the theories which were the most blame-removing, and so on. . . . The patients became catatonic again, and we proceeded from there. Some of them, praise God, have been well enough to be out of the hospital for a good many years now.[42]

Sullivan's conception of the forms of schizophrenia as representing different, dynamic aspects of one and the same total process, explains his *ambivalence* toward Kraepelin's system. On the one hand, Sullivan accepts, of course, the standard Kraepelinian classification of schizophrenia into catatonic, paranoid, and hebephrenic forms: this is a foundational explanatory grid. On the other hand, he rejects Kraepelin's system because Kraepelin, according to Sullivan, did not see clearly that these are all different aspects or moments of one and the same "disease entity" or "disease process." The person with schizophrenia can easily transition between these forms, and the Kraepelinian system artificially severs them:

I believe that Kraepelin was unfortunately driven by a certain obsessional necessity for completeness which made naming things and juggling with nosological entities one of the most superior forms of human activity, which I think was probably an attribute of a certain type of German intellectual. Thus even though he made the great central dementia praecox synthesis, he failed to notice that these things that come very near being separate diseases are rather strikingly functions of the personality structure concerned and that such characteristics as age of onset, and so on, point in that direction rather than in the direction of almost separate entities. But that would have been awkward. That would have been recondite and unclear. And so we got dementia praecox.[43]

Sullivan's work begins the process of drawing out a theme that was latent in Heinroth: the idea of the healer as guide and shepherd. For Sullivan understands and accepts, as does Heinroth, that some forms of madness can be advantageous; they represent a healing journey, a "wholesome desire of nature to cure a perversion through another perversion," and that other forms of madness represent aberrations, or ways that the healing journey can be aborted. In that, Sullivan and Heinroth are one. But Heinroth, the physician, never fully insets *himself* into this picture. In other words, the healer, for Heinroth, does not have a special role to perform in this process. The healer is not there to ensure that the patient remain on the true path; his job is not to encourage the patient or to mark out or highlight the true path. Heinroth simply describes the various ways that this form of madness can go wrong. He doesn't seem to see any immediate consequences for his own role as a healer. But Sullivan does. Sullivan understands, however dimly, that his role as healer is to encourage the person to remain in the catatonic state as long as possible and not to accept the cheap paranoid or hebephrenic "solutions." As noted above, he takes evident pride in the fact that, on a handful of occasions, he has been able to successfully "push" the paranoid person back into the catatonic state. But this theme of psychiatrist as shepherd and guide only becomes fully explicit in Laing.

Philippe Pinel, like Heinroth, understood that, in some cases, madness has a therapeutic function; the purpose of the healer is *not* to disrupt or stifle the psychotic episode, or to bring it to a premature halt. Psychotic episodes, what he called *accès de Manie*, are healthy, not because they represent a needed flight from reality, that is, a coping strategy, but because they are *physiologically invigorating*. They are nature's antidote to a poisonous lethargy that has gripped the mind.

9

A Salutary Effort of Nature

Philippe Pinel, one of the founders of the "moral treatment" of the mad, begins his 1800 treatise by describing one particular form of madness at length, *manie périodique ou intermittente*, characterized by intermittent attacks of fury. The fact that Pinel spends so much time describing this disease suggests that, for Pinel, *manie périodique* is somehow paradigmatic of *all* madness whatsoever (Section I). Ultimately what we discover is that *manie périodique* is distinctive in its teleological and goal-directed character. These attacks of madness, he argues, have a cathartic or "salutary" effect on the patient. It is for this reason that, as a rule, they ought not be medicated but be allowed to run their course (Section II). Finally, we must ask what, precisely, the relationship is between his teleological conception of madness and his moral treatment. Moral treatment, it turns out, is a natural consequence of adopting a teleological view of madness: if some forms of madness are by design, then the goal of the doctor should simply be to create a safe and non-threatening environment for them to run their course as nature intended (Section III). Hence madness-as-strategy has profound implications for models of treatment.

I.

Pinel describes a group of five young men admitted to the Bicêtre "with a sort of obliteration of the faculties of the understanding or what we can name a dementia of imbecility."[1] Each continued in this state from three months to over a year. Eventually, each one had a bout of mania (*accès de Manie*) characterized by formidable violence; *immediately thereafter*, "all these madmen [*insensés*] recovered the use of their reason."[2] These attacks of mania, these maelstroms of fury, violence, and apparent psychic pandemonium, *were part of the healing process.*

Given the salutary effect of these attacks, should we not, then, think of the one who wants to medicate these madmen, to "cure" their attacks, as himself

Madness. Justin Garson, Oxford University Press. © Oxford University Press 2022.
DOI: 10.1093/oso/9780197613832.003.0010

mad? *The true madness*, Pinel concludes, would be the attempt to stifle their occurrence: "I ask now if all physicians who seek to 'cure' similar attacks, do not deserve to be put in the place of the madman himself?"[3] In case there is any doubt as to who Pinel thinks is truly insane, doctor or patient, he later observes that "bleeding can sometimes be lavished with so little discernment, that one can almost put into doubt which of the two is the most insane, the one on which it is practiced or the one who orders it."[4]

With this, Pinel places the study of madness on a ground that is quite foreign to the ground that we have trod so far. He situates madness in a fundamentally *non-Kantian space*. What we must encounter in Pinel is not, in the first place, the moral reformer who, with revolutionary zeal, broke off the chains of the mad patients at the Bicêtre; nor must we think of Pinel, first and foremost, in his role of patriarch who thought that in order for the mad to be healed, in order for their unreason to be transformed into reason, they had to submit themselves, body and mind, to the reason of the hospital's governor, as a small child does with her father.[5] Rather, what must come to mind, first and foremost, is his rethinking of the relationship between madness and teleology in a non-Kantian reference frame. And when we unearth this altered relationship between madness and teleology, all of those other things—the moral treatment, the breaking off of the chains, the paternalism—will be seen as natural, or at least comprehensible, consequences of this new relationship.

Pinel's *Traité médico-philosophique sur l'aliénation mentale ou la manie* begins with a very lengthy deliberation on a specific form of madness. This is periodical mania (*manie périodique ou intermittente*), characterized in the following way in the preface:

> Intermittent or periodical insanity is the most common form of the disease, and the deviations in the understanding which characterize these attacks, correspond to those of continual mania, and give an accurate idea of it.... It is therefore by their historical exposition that this treatise should begin; the principles of moral treatment should follow immediately after, since often enough it alone can bring about healing, and since if one neglects it, these bouts of mania are exasperated, they become more obstinate or are converted into a continuous and incurable mania. This sort of moral institution of the mad, proper to ensure the reestablishment of reason, assumes that in the greatest number of cases there is no organic lesion of the brain or the skull.[6]

The fact that he begins his treatise with this *particular* form of madness, *manie périodique*, should give us pause. First, *why* begin the treatise with a lengthy description of one particular form of madness? After all, Pinel has a lot of ground to cover in his treatise. He must introduce the principles of moral treatment. He must provide a nosology. He must discuss medications, which remedies to use and when to omit them, and so forth. So, it is a puzzle that Pinel decides to spend the first part of the book—roughly, the first 50 pages of a 300-page treatise—by discussing one peculiar form of insanity. Why begin a medical text with a lengthy deliberation on the nature of this style of insanity, when there is so much work to be done, so much labor ahead? Historians of Pinel seem to have overlooked the question entirely.

True, *manie périodique* is the most *common* form of insanity; perhaps this is the justification for devoting so much of the text to it? Relative frequency alone, however, hardly justifies allotting so much space to it in a book that is intended as an overview of *all* forms of madness. What then, if not that *manie périodique* is an *exemplar* of all madness, a paradigm of *all* of its forms (*mélancolie, manie sans délire, manie avec délire, démence, idiotisme . . .*)? Perhaps there are core features of it *thorough reflection on which* will allow us to draw out features of all madness, will allow us to capture its elusive essence? The starting point of such a self-consciously momentous text is not arbitrary. He wishes to orient us in this nascent field of psychiatry.

There is a second rationale for this selection. Not only will the study of *manie périodique* reveal the elusive essence of madness, but it will set the stage for the basic methodology of healing, the *traitement moral*. In other words, once we understand what *manie périodique* is, we will be able to deduce some consequences for healing it, and, by extension, for healing madness generally. So, there is both a theoretical and a practical strategy being deployed here. Moreover, Pinel, in the passage above, tells us that the fact that the principles of moral treatment are, *as a rule*, successful, implies that, *as a rule*, madness does not stem from an organic lesion in the brain. This is an extraordinary claim. It would be as if I announced, "I have discovered a treatment for panic disorder, and I have found that my treatment is applicable to most forms of mental illness. Consequently, since my treatment does not involve physical means, I do not think mental illness, *in general and as a rule*, has a physical seat in the brain." *That*, if true, would justify devoting a very long opening section to one particular illness (say, panic disorder) in a treatise on mental illness generally.

What, then, is so distinctive about *manie périodique*? As the name implies, a patient with this disease undergoes intermittent bouts of mania (*accès de manie*): delusions, hallucinations, often accompanied by violent fury. These attacks generally have an abrupt beginning and end, they continue for three to six months out of the year, and are often induced by the summer heat. What makes this the window onto madness in toto?

On the way toward an answer to that question, we must take note of a quite distinctive feature of *manie périodique*: its *reasonableness*, and this in two senses. The first is a diachronic sense: a person with *manie périodique* might be entirely "in their senses" for six to nine months out of the year. We are dealing, here, with someone who is *generally a reasonable person*, someone who can be reasoned with, someone who appears, for the majority of the time, to be perfectly sane, who can, during these extensive lucid intervals, re-flect on his condition and even detest these periodic bouts of insanity.

The condition also announces its reasonableness in a second sense, not only in the sense that, diachronically, reason alternates with madness, but that for some patients, even *during* these periods of mania, *reason continues to operate in a completely unimpaired manner*. There are a minority of such patients, Pinel tells us, who retain the functions of the understanding but who simply find themselves with an overwhelming urge to destroy eve-rything around them. These individuals will even alert their friends and family when they feel one of these attacks coming on so that they do not hurt their loved ones. He mentions three such cases with the following char-acteristic: "These madmen responded in a manner most accurate and pre-cise to the questions that one proposed to them, but they were dominated by a most spirited fury, and by a bloodthirsty instinct, of which they themselves felt horrified."[7]

It is, in fact, on the basis of such exceptional individuals that Pinel rejects the universality of Locke's formula that the madman "reasons correctly from false premises." No doubt it is true that madness sometimes emerges from a misassociation in an otherwise sound reasoning ability. But in some patients, Pinel emphasizes, as does his student Esquirol, there is no *intellectual* failure at all, whether in the association of the ideas or their manipulation, neither in the "grist" nor the "mills" of thought.[8] The trouble stems only from the in-tensity of emotion.

The fact that madness can occur alongside a perfectly intact reasoning power forces us to confront the issue of whether madness, as in Kant, or in Haslam, always and everywhere results from a failure of some kind of faculty of the mind to perform its office, such as imagination. Pinel's answer is am-bivalent. As a rule, for Pinel, it does: we can generally trace the disorder to a perversion or diminution of a faculty of the mind, whether it be reason, im-agination, or will:

> Sometimes these functions are altogether abolished, weakened, or deeply excited during these bouts; sometimes this alteration or perversion falls only on one or many of them, while the others have acquired a new degree of development and activity which seems to exclude all idea of alienation of the understanding.[9]

That said, Pinel also emphasizes, to a degree that Kant and Haslam do not, the extent to which madness can be associated with the almost limitless aug-mentation of a faculty, in a manner worthy of our encomium:

> One day I asked of one, with a very cultivated mind, to write a letter to me at the very moment at which he held the most absurd propositions, and yet this letter, which I still have, is full of sense and reason. A goldsmith, who had the extravagance to believe that his head had been switched, be-came obsessed at the same time with the illusion of perpetual motion; he obtained his tools, and he set himself to work with the greatest obstinacy.... [I]t resulted in very ingenious machines, the necessary fruit of the deepest combinations.[10]

And even further:

> In some cases, this reaction of the epigastric forces on the functions of the understanding, far from oppressing or obscuring them, actually augments their vivacity and energy. . . . The attacks seem to carry the imagination to the highest degree of development and fecundity, without ceasing to be reg-ulated and directed by good taste. The most remarkable thoughts, the most ingenious and exciting connections, give to the mad a supernatural air of inspiration and enthusiasm. The memory of the past seems to unroll with ease, and what had been forgotten in the calm intervals is reproduced now in his mind with the most lively and animated colors.[11]

Though Pinel recognizes that madness sometimes results from the break-down of a faculty of the mind, it can also involve the augmentation of one or more faculties; because of this, the mad person's ruminations and inventions can be worthy of our utmost respect, even our quiet awe.

Back to *manie périodique*. This condition is marked by its *reasonableness*. The fact that Pinel spends so much time at the outset of his treatise describing this condition suggests that he thinks of it as paradigmatic of madness as such, that is, that *a certain degree of reasonableness is inherent in madness*. But by this point, we should not be surprised to find reason in madness: Kant, Haslam, Wigan, and Heinroth all struggled to make sense of the same puzzling observation. In fact, the whole basis of Pinel's moral treatment rests, he says, precisely on the fact that the mad person *yet possesses a measure of reason*. Pinel approvingly quotes a passage from an *Encyclopaedia Britannica* entry on the subject: "In the moral treatment . . . one does not consider the mad as absolutely deprived of reason, that is to say, as inaccessible to motives of fear, hope, the feelings of honor. . . . One must subjugate them at first, then encourage them."[12] We have yet to understand, however, the precise role of *manie périodique* in prompting the return to sanity.

As an aside—in particular, regarding the idea that the governor must "subjugate," then "encourage"—we should recall Foucault, who famously wrote that, though Pinel and Tuke, through their moral therapy, broke off the chains of the mad, they also instituted a different and perhaps more insidious form of violence; they recast the mad as children who were in need of discipline and order.[13] Without doubt, there is some truth in Foucault's characterization: Pinel thought that his mad charges required a kind of *surrogate* reason and that the governor's role at the Bicêtre was to provide that surrogate reason, and Tuke is even more explicit that the relation between patient and governor must mirror the relation between child and parent. But we must be very clear that this arrangement does not imply that this patriarch/governor is a surrogate reason, *as if the mad had no reason of their own*. Rather, this arrangement is effective only to the extent that the mad are possessed by a certain *measure* of reason, as the *Encyclopaedia Britannica* passage notes well. That is because the madman's very ability to submit to another person's reason *as* reason, his ability to say, "I see you, your thoughts, your words, your actions, your temperament, your gentle but firm hand, as a kind of pinnacle of reason, as something that I aspire to enjoy for myself," his ability to make that very judgment, is to be, to that very extent, reasonable. Moral treatment works because it seeks out and latches onto the madman's intrinsic reasonableness. And it is this intrinsic reasonableness that the treatment attempts to draw out and fortify. For example, Pinel tells us of a mad soldier who, for a week, was gripped by furious anger to the point of requiring restraints:

> Eight days passed in this violent state, and he finally seemed to understand that he is not the master of his own whims [*il n'est pas le maître de suivre ses caprices*]. During the governor's morning rounds [*la ronde du chef*], he [the soldier] took the most submissive tone, kissing his hand: "You promised me," he said, "to set me free inside the hospital if I was calm; and I ask you to keep your word." The other, smiling, expressed the pleasure he felt at this happy return to himself [*cet heureux retour sur lui-même*]; he spoke to him with tenderness, and in that instant, he removed the constraints, which would have been superfluous or even harmful.[14]

Clearly, the governor, here, adopts the archetype of a powerful father—in this, Foucault is correct—but this arrangement does not work by ignoring or denying the madman's reason but *affirming* it: here, it is the madman's own recognition that *he is not is own master* that constitutes the turning point in the treatment. At that point, the patient does not become a new person; he "*returns to himself.*"

II.

We still have not yet discovered the answer we seek. Why does Pinel begin his treatise with *manie périodique*? We should not rest content with this correct, but rather trite, formulation: "Pinel begins his treatise with *manie périodique* because *manie périodique* exhibits a fundamental reasonableness which is paradigmatic of all madness and which sets the stage for his moral treatment." That would be correct, as far as it goes, but it remains superficial, because it does not yet explain the role of the attack of mania in facilitating sanity, *its mechanism*. It does not explain, yet, why the physician who wishes to stem the attack of madness is himself, quite possibly, the one who is truly mad. It does not explain why Pinel contravenes the current medical wisdom that insists the attacks *must be stifled*.

And, in fact, the idea that these attacks *must be stifled* is a common medical belief. Haslam, for example, is a staunch advocate of the idea that these attacks must be stopped as quickly and as early as possible, as part of the healing process. That is because, for Haslam, madness is a disease of misassociation, as Locke said. During these flights of mania, the mad person is extremely susceptible to forming new associations, adding to the perverse mesh-work of associations that constitutes his madness. Thus, any stimulus that he encounters will be assimilated by this growing beast, this organism, this body of associations; it will be immediately subsumed within it. I believe that the CIA is after me; in my flight of madness I see one doctor whisper to another; she becomes part of the fabric of delusions; she is now part of the conspiracy, a new thread in the meshwork of association. The fact that the attack of mania has a self-reinforcing and self-perpetuating quality is the reason it must be brought to a swift end.

For Haslam, there are two ways we can stifle these attacks. First, to halt the "raving paroxysm," the patient should be bled profusely—but better to draw blood from the scalp than from the arm, for "by these means any quantity of blood may be taken, and in as short a time, as by an orifice made in a vein by the lancet."[15] Second, we can put the furious patient in restraints in a dark room, so that no new stimulus can work its way into this growing machine, the growing monster of association:

> In the most violent state of the disease, the patient should be kept alone in a dark and quiet room, so that he may not be affected by the stimuli of light or sound. . . . As in this violent state there is a strong propensity to associate ideas, it is particularly important to prevent the accession of such as might be transmitted through the medium of the senses.[16]

According to Pinel, Haslam is, in this regard, *the true madman*. Haslam him-self is mad for wishing to stifle these attacks. Yet we still do not yet under-stand what, precisely, is so mad about Haslam's method. We understand that these attacks seem to provide some relief; they seem to perform a mysterious cathartic function; but we do not yet understand the *how* of it. What is its mechanism?

Chapter 13 of the first section of Pinel's *Traité* is where he deals squarely with the healing properties of madness and its underlying mechanism. It has the following unusual title: "Motives for regarding most bouts of mania as the effect of a salutary and favorable reaction for healing." We will read the chapter slowly, for there is much to unpack. It begins:

> We love to soar with Stahl above this philo-pharmaceutical medicine, bris-tling with its formulas and its small means, and to rise, even regarding

mania, to the general consideration of a conservative principle [*un principe conservateur*] which seeks to repel all harmful attacks by a happily combined series of efforts, just as in fever.[17]

The passage begins with an invocation of the early eighteenth-century German physician and chemist Georg Stahl. Stahl's view of fever famously broke with the traditional Greek view that fever is part of the disease, or the very disease itself; a "heat contrary to nature," as Avicenna described it. In contrast, for Stahl, picking up on the Greek notion of the *vis conservatrix et medicatrix naturae*, fever is part of the organism's attempt to heal itself. It is salutary; it must run its course for healing to occur.[18] Nature is infused with a secret vitality. Nature is not a blind mechanism; it is a goal-directed, goal-driven process that tends toward healing. It is for this reason that Stahl is among the patron saints of the contemporary naturopathic movement; moreover, as Bynum points out, Stahl's view anticipates, by roughly three centuries, the doctrines that constitute so-called Darwinian medicine, according to which many of the symptoms that we have traditionally considered to be part of the disease, as part of the pathology, are in fact adaptations, such as morning sickness in first-trimester pregnancy, often construed now as evolution's strategy for preventing toxins from infiltrating the fetus.[19] In Pinel's invocation of Stahl, we find ourselves having crossed over to a fundamentally new position for understanding these bouts of mania and for treating them. The extraordinary thought that Pinel entertains, here, is whether, and to what extent, *what is true of fever can be true of insanity.* As he asks in the passage above, does nature's principle of conservation apply "*even regarding mania*"?

Let's continue the passage. Pinel informs us that these bouts of mania are precipitated by disturbances in the gastric region; the attack of mania then helps to bring these disturbances to an end. These maniacal attacks are in some yet to be fully understood manner part of the healing process:

A lively affection, or, to speak more generally, any stimulus whatsoever, acts forcefully on the center of the epigastric forces (V), and produces a deep commotion which repeats itself on the abdominal plexus, and gives rise to a spasmodic constriction, to an obstinate constipation, to inner heat (VI). Soon after, it excites a general reaction, more or less strong depending on the individual sensibility; the face is colored, the circulation becomes more lively; the center of the epigastric forces seems to receive a secondary impulse of a completely different nature than the primitive one (V), the muscular contraction is full of energy; it often excites a blind fury and uncontrollable agitation; *the understanding itself is drawn into this sort of salutary and combined movement* (VII).[20]

At least at the outset, we are not witnessing a "mental disorder." The disease, in the first place, is not a breakdown in a *mental* function. It is an abdominal disturbance, and the *fury*, the *agitation*, and even the *understanding* are somehow drawn into the process that ultimately produces healing:

> These [mental] functions are altered, or many at a time, and sometimes they redouble their vivacity. It is in the midst of this tumultuous trouble that the gastric or abdominal affections cease, after a more or less prolonged duration (X); a calm follows, and brings a healing all the more solid as the attack was violent.[21]

The attack of mania, then, plays a role parallel to that of fever in Stahl's theory. It is a healing and salutary movement of the mind; the changes in the functions of the understanding, the delusions, the ravings, the hallucinations, are somehow or another part of this "movement." And this "movement" is not by accident, it is by design, but *nature* is the designer: nature has established this strict sequence of events as part of the recovery process.

Pinel illustrates this lofty principle by relating the story of the five young men who were admitted to the Bicêtre with "obliterated" faculties and who were all healed when their attacks were allowed to run their courses. He concludes by posing the question that, at this point, does not seem entirely rhetorical: "I ask now if all physicians who seek to 'cure' similar attacks, do not deserve to be put in the place of the madman himself?"[22]

It remains to be seen, however, by what secret power, by what vitality, these attacks possess their healing virtue. We can venture the suggestion that, for Pinel, these attacks have an *invigorating* quality, and that by stirring up the blood they literally induce a catharsis—a purging—of the source of the disturbance. He returns to this theme of the salutary quality of these manic attacks throughout the *Traité*. One passage is particularly illuminating. He tells us that these *accès de Manie* can cure melancholy *and even idiocy*:

> Some, mostly in their youth, after remaining for several months, or even entire years, in an absolute idiotism, fell into a sort of bout of mania that

lasted 20, 25, or 30 days, and each was followed by the reestablishment of reason, by a sort of internal reaction.[23]

Here, Pinel discloses something like a *mechanism*. For the problem, in *idiocy*, is an ebbing of the vital powers and faculties of the mind, a state of "weakness, atony, and stupor."[24] The attack of mania must somehow work by reversing this trend toward stupor; hence, by *quickening* the mental powers.

The idea that these attacks of mania have a purpose, a goal, a telos, has implications for Pinel's ethos of non-intervention. As the classicist Ludwig Edelstein observed, "In later centuries [following antiquity] the recognition of nature as a teleological power must have confirmed the advisability of the withdrawal of the physicians."[25] This is certainly true of Pinel, who tells us, at the very outset of the *Traité*, that "in mania, as in many other maladies, if there is an art for administering medications, there is an even greater art in knowing when to do without them."[26]

Pinel's student Esquirol carried on this tradition of *non-intervention*, particularly in the case of fury (*la fureur*); he held fast to the same mechanism, namely, that fury somehow reverses the ebbing of the vital energy: "Furious maniacs and monomaniacs, are cured more frequently than those who are calm and easy. In the latter, there is less force, less energy, less reaction; cures are more difficult."[27]

III.

We have arrived at our goal. What makes *manie périodique* an appropriate starting point for a treatise on madness? It is that it reveals something of the essence of madness while, at the same time, pointing the way toward treatment. What these *accès de manie* reveal is the teleological character of madness; this character is part of its essence. With Pinel, we are continuing to rethink madness in a non-Kantian manner. But what is worth remarking upon is that, in contrast to the Middle Ages and even the Renaissance, this telos is no longer tethered to theology, to the dynamic of sin, punishment, redemption. Pinel's orientation is broadly naturalistic. The healing tendency of nature is not some kind of supernatural principle added on top of nature; Pinel is not advocating a new form of dualism. Rather, it is simply a way of conceiving of nature itself. *Nature itself possesses an intrinsic purposefulness; it does not need to be given this purposiveness from an "outside."* It remains for other thinkers, for future thinkers, particularly Kurt Goldstein and the evolutionary psychologists of today, to demonstrate how madness can be perfectly teleological and perfectly natural, rather than "natural to the extent that it rejects teleology," and vice versa.

We have understood the teleological nature of *manie périodique*, but it remains to be seen how this perspective gives rise to the moral treatment for which Pinel is so famous. Part of the received historical wisdom about Pinel is that his breaking off of the chains of the patients at the Bicêtre has a two-fold root: the first is a new, liberatory ethos wrought by the French Revolution; the second is his receipt of accounts of the success of moral treatment in other places, likely including accounts of William Tuke's York Retreat.[28] So, there is an ethical and political drive toward liberating the mad, which is inconsistent with restraints, and there is also solid clinical and empirical justification for this shift in practices. And those things might be true, but there is a third source, and that is Pinel's altered vision of the relationship between teleology and madness. For chaining up the madman, pummeling him with medications, throwing him into a dark room, is effectively disrupting this purposeful movement, this *vis conservatrix et medicatrix naturae*; it is preventing madness from reaching its zenith and end. Instead, he says, the role of the doctor is to allow it to reach its natural end, to use restraints only if absolutely necessary to protect the safety of the sufferer or those around him, and for as brief a time as possible; to use medications only in the rare case in which the series of such attacks yields no healing fruit:

> The constant laws of the animal economy considered with respect to mania as in the other maladies, fill me with admiration for their uniformity, and I see again the unexpected resources of nature left to itself, or wisely directed; this has made me more and more frugal regarding the use of medication, that I have stopped using it, until I have convinced myself of the insufficiency of the moral remedies.[29]

It is only because madness possesses this future orientation, because *madness is trying to do something*, that the art of healing must consist in remaining on "standby": allowing it to happen while curbing any deviations in its trajectory. As we have noted, Pinel emphasizes how damaging, even fatal, it can be when the healer attempts to disrupt these attacks because she sees in them mere pathology, nothing more than the result of a breakdown of some mental function: *the blood of maniacs is sometimes so lavishly spilled . . .*

Furthermore, not only does he warn us about what a fatal mistake it can be to see these attacks as pathological rather than as having a healing function, but he also says that the very idea that madness stems from an organic brain dysfunction is itself responsible for the persistent pattern of abuse and neglect in the asylum.

> A general and very natural opinion is that the alienation of the functions of the understanding consist in a change or a lesion of some part of the head. . . . [F]rom this comes the prejudice of regarding this malady as most often incurable, to simply sequester the mad from society, and to refuse them even the help that all infirmities deserve.[30]

Here we see yet one more distinction between Haslam and Pinel, for Haslam, perfectly aware of the extent of abuse, cruelty, and inhumanity in the asylum, public or private, sees it as merely an accident, as an effect of poor hiring practices, as a result of not properly vetting the managers and guards. For Pinel, the problem is structural; it is something expectable; it has an ideological reason or ground. We might then pause when we hear mental health advocacy groups insist that treating madness as a brain dysfunction is going to eliminate cruelty or stigma against the mad. It is possible that, at least in the early nineteenth century, it had quite the opposite effect.

10

The Biologization of Kant

Wilhelm Griesinger, in his *Die Pathologie und Therapie der psychischen Krankheiten* of 1845, embraces a Kantian view of madness while *biologizing* it, that is, while rethinking it in the space of brain organs and functions. Now, mental disorders are dysfunctions of the nervous system. To know how many basic kinds of mental disorders there are, one must know how many parts of the nervous system there are and what their jobs are (Section I). Still, despite his dysteleological starting point, Griesinger is unable to relinquish teleology entirely. His basic understanding of delusion is that *delusions, like dreams, are wish fulfillments.* They are forms of refuge against terrifying or traumatic experiences (Section II). The task of finally purging psychiatry of teleology was left to the great nosologist Emil Kraepelin. For Kraepelin, there cannot possibly be teleology in madness simply because, to the extent that we can glimpse purposiveness and goal-directedness in the mad, *we are not really dealing with madness*, but malingering (Section III). For better or for worse, Kraepelin's accomplishment—to rid psychiatry of teleology—was overshadowed by psychoanalysis, which places the study of *all* madness onto a firm teleological footing.

I.

Kant's vision of madness operates in the abstract space of mental faculties and their functions. There are three main faculties—judgment, experience, reason—each of which performs a designated role in the economy of the mind. It is a clean space; it is a hygienic space. But how does this sterile space relate to the physical space of the brain, to the gray and white matter in the skull? In his "Essay," Kant gestures, in a passing way, toward the possibility of an intimate relationship between the two spaces: "I have also [in this essay] only paid attention to their [that is, the maladies of the head] appearances in the mind without wanting to scout out their roots, which may well lie in the body and indeed may have their main seat more in the intestines than in the

Madness. Justin Garson, Oxford University Press. © Oxford University Press 2022.
DOI: 10.1093/oso/9780197613832.003.0011

brain."[1] But his speculation that bodily dysfunction is implicated in madness is never raised to the level of a serious, sober commitment. Not only is he non-committal with regard to *whether* madness has its roots in the body, but also, if it does, the region of the body it stems from, the brain or the intestines. With Kant, then, we have a *layered* ontology; we have the space of the bodily states, processes and functions, and we have the space of the mental faculties and their functions. We can remain neutral about how these spaces are tied together, and the lines that connect them.

With Griesinger, this layered ontology is flattened. The faculties of the mind are *compressed* into the states and processes of the brain. The functions of the mind's faculties (memory, imagination, will . . .) are functions of the brain's organs. It is not as if there are two *sorts* of entities standing in a complicated relationship: the brain states and the mental faculties. There is simply one entity: the brain, embedded in the nervous system as a whole. While Griesinger freely admits that, as a point of logic, he cannot *demonstrate* the truth of this materialist hypothesis, he assures us that the materialist point of view is the only perspective that a scientific psychiatry need adopt:

> How a material physical act in the nerve fibres or cells can be converted into an idea, an act of consciousness, is absolutely incomprehensible; indeed, we are utterly unable even to settle the question of the existence or nature of the media existing between them. All these matters are as yet only probable; in which state of affairs the simplest hypothesis is the best; and certainly the materialistic offers fewer difficulties, obscurities, and contradictions than any other, especially in relation to the origin of thought.[2]

Still, Griesinger's project, though it flattens Kant's layered ontology, remains, in essence, Kantian. He retains as axiomatic that *there are as many basic forms of insanity as there are faculties of the mind, and that is because each form of insanity is simply an aberration of the corresponding faculty*. The only difference is that these faculties are now brain states and processes. An affliction of a faculty of mind *is just* a pathology of the corresponding state or process in the nervous system.

In effect, Griesinger urges that psychiatry undergo *two* quite distinct transformations operating in parallel, though he himself probably did not think them through clearly. *First*, madness should undergo a transition from the abstract space of mental faculties to a biological realm of brain organs. *Second*, madness should undergo a transition from teleology—as in

Heinroth, Wigan, Pinel—to dysteleology. The fusing together of these two doctrines—that madness is biological, and that madness is dysfunctional, or more concisely, that *mental disorders are biological dysfunctions*—defines the era that historians call "German imperial psychiatry."[3] These twin transformations outline an intellectual space for understanding what took place in German psychiatry from the period that begins, roughly, in 1845, with the publication of the first edition of Griesinger's *Die Pathologie und Therapie der psychischen Krankheiten*, and that ends, roughly, with the 1899 publication of the sixth edition of Kraepelin's textbook and, in the same year, with Freud's *The Interpretation of Dreams*.

The term "German imperial psychiatry" points, first and foremost, to a metaphysics of madness, and only secondarily to a historical era. This Kantian project, this dysfunction-centered project, begins with a picture of the sane mind. Such a project must, logically even if not chronologically, start with a survey of the faculties of the sound mind, even if this survey remains implicit and rudimentary. If we are to biologize this Kantian project, then, it is imperative to begin with a basic understanding of what the nervous system is *about*, what the nervous system is trying to do, what its natural parts are, and how these parts relate to one another and to the life of the organism.

Let us peer into the nervous system as a whole. At the outset, we find two major components: the spinal cord and the brain. But the nerve tissue inside the spinal cord itself divides along two pathways, what Griesinger calls the centripetal pathway and the centrifugal pathway, so the nervous system has three main parts in total: the centripetal pathway, the brain proper, and the centrifugal pathway, corresponding to what we now call the "flow of information" from stimulus to behavior. The centripetal pathway is the one that carries the external impressions to the brain. Importantly, the centripetal pathway *does not just passively reproduce sensations*; it has its own formative work to accomplish. One of its jobs is to somehow transform the pattern of irritation of the end organ in such a way that it can be utilized by the brain, in a format or currency that the brain can exploit: "Towards the brain it conducts certain qualities of sensation . . . which evidently also originate in the grey substance [that is, the spinal nerves] itself, and constitute a sort of 'psychical' modification and transformation of the centripetal impressions."[4] At the opposite end, the centrifugal pathway converts the brain's commands into bodily movements. It, too, imposes its own form on the brain's commands; it translates them into a format the body can recognize and act upon: "Inversely, the impulses of movement from the brain do not appear as yet to possess all the qualities necessary to isolated muscular contractions. It appears to be in the grey substance [the spinal nerves] that these impulses are first elaborated and arranged in a proper manner."[5]

We know what the centripetal and centrifugal pathways do. But what does the brain do? What is *its* function? Simply put, for Griesinger, the purpose of the brain is to compile something like (as we would now say) a *library of sensorimotor contingencies*. Its job is to figure out which motor output is to be performed in response to which sensory input, or in other words, to assign certain sensory stimuli to certain motor outputs: "But the brain is also an immense reflex apparatus, in which all these states of sensorial excitation, of which this organ is almost constantly the seat, are transformed into impulses of movement."[6] But the fact that the function of the brain is to compile such a "library," the fact that there even needs to be some faculty, some device, that works out the right mapping, raises a deeper question: *why do we need a brain at all*? Why must *anything* perform this function? After all, simple transformations of sensory input into motor output do not always require the mediation of the brain: the "principal function" of the spinal marrow is to produce "the more simple reflex acts, the transformation, still pretty direct, of sensations into movements."[7] The knee-jerk reflex, as we are often told, is exclusively "mediated by" the spinal cord. Though Griesinger does not answer that question directly, he recognizes that the biological justification for having a brain *must be*, in the final analysis, to facilitate the process of converting sensory irritations into bodily movements. *The job of the brain is not, in the first place, to think*. True, in order for the brain to do its job, it requires subsidiary functions—memory, imagination, abstraction, motivation, an *I*—but all of those functions are collapsed under, subordinate to, this one fundamental goal, which is sensorimotor adjustment.

The mind is encapsulated; it is simply one among many cogs in a complex machine that is up to the business of surviving and reproducing. To know the human being is not to know the mind, in the first place; rather, it is to know how this thing called mind contributes to the business of eating, breeding, breathing, fighting, making a living. Arguably, this is *all that it is* to take a biological approach to the mind: to see the mind in its aspect of adjusting sensorimotor contingencies as a way of perpetuating the life of the individual, and ultimately, the species.

There are, therefore, only three elementary ways that the mind can fail, that is, three basic forms of mental disease. It can fail in the sensorial function; it can fail in the motor function; it can fail in this intermediate function, what we traditionally call "mind":

> In those cerebral affections which come under consideration as mental diseases, there are, as in all others, only three essentially distinct groups— namely, sensitive, motor, and mental (perceptive) anomalies. Thus, according to this threefold division, we have to consider successively each of the three leading groups of elementary disturbances—intellectual insanity, emotional insanity, and insanity of movement.[8]

It is true that these are only the most general categories; they are what Kant calls the *Hauptgattungen* and what Griesinger calls the elementary disturbances (*Elementarstörungen*); they are *elementary* in that they are the most general, and all of the others are either disorders in subordinate functions or hybrids of these elementary ones.

II.

At the most general level, there are three different ways that something can go wrong in the mind, that is, the nervous system. But what is it for, say, the sensory apparatus to "go wrong"? What is it for the motor apparatus to "go wrong"? It seems we already know the answer: it is for there to be some sort of pathology or lesion or degeneration or obstruction: "shock, injury, excessive fatigue . . . excessive mental irritation through emotion and the like . . . other further morbid changes in the organism . . . anomalies in the circulation . . . deficient nutrition and excitation of the brain."[9] But there is a metaphysical problem here: what is the *norm* under which we would judge a certain lesion to be pathological? What makes a change in the organism a *morbid* change? What is the basic rule or standard, *what is the nervous system supposed to be doing*, such that *certain* kinds of lesions, *certain* kinds of obstructions, *certain* kinds of anomalies constitute normal variation, and other kinds of lesions, other kinds of obstructions, other kinds of anomalies constitute pathological variation?[10]

At the most general level, in order for this complex sensorimotor apparatus to contribute to its basic task of survival and reproduction, it must be properly oriented toward the world. It has to be *calibrated* to the world; it must be responsive to the world, under the sway of the world:

The essential characteristic of insanity, that which distinguishes it as a morbid state, consists chiefly in the fact that certain states of the brain—certain dispositions, feelings, emotions, opinions, determinations—proceed from within outwards, owing to disease of the organ of the mind; while, in the healthy state, our emotions, opinions, determinations, originate only upon sufficient external motives, and on that account also stand in a certain harmonious relation to the external world. No one wonders if any one who had sustained a great loss is sad, if another to whom an earnest wish has been fulfilled manifests exuberant joy; but we rightly consider it as a morbid symptom when an individual without any external motive is depressed with sadness or elated with joy, or even, where some external cause has been given, the individual is immoderately or for too long a time affected by it, as when a trifling occurrence excites an individual to violent rage which he is unable for a long time to restrain.[11]

It is easy to see that the sensory system "goes wrong" when it produces hallucinations. But why? Because the sensory system is acting in a manner that is autonomous from the world; it is not properly calibrated or subordinated to the world. Suppose brain damage inside of me causes the perception of a floating dagger, but there is no floating dagger. My brain activity is not properly *responsive* to the world; therefore, this is a form of insanity. But that same rule, that something in the mental apparatus is disordered when it is not properly under the sway of the outside world, is just as true for our emotions, for our thoughts, and for our motivations, as it is for our perceptions. If I see, in the distance, a man approaching me with a dagger, and I see him truly and veridically—there actually is a man with a dagger—and I feel, not fear, but hilarity, or if I form the idea that the dagger cannot hurt me because I have a magic stone, or if I feel fear and I know that the dagger can hurt me but I am nonetheless gripped by the urge to run to him, then we can say that my emotions, my thoughts, or my motivations are, respectively, disordered, because they are not properly under the sway of the world.

So far, this is what the nervous system is about: sensory impressions are transformed by the brain into motor impulses. The function of the brain is to mediate this transformation and, over time, to compile a library of sensorimotor contingencies. But where are the faculties of the mind? Where is perception, imagination, emotion, reasoning, will? Where is the *I*? These faculties are not housed in different portions of the brain but all of them are somehow subsidiary functions to this main function of transforming impressions into impulses:

> All mental acts take place within the intelligence. This is the special seat of thought, and all of the various mental acts that were formerly designated separated faculties (imagination, will, emotions, etc.) are only different relations of the understanding with sensation and movement, or the result of the conflicts of ideas with themselves.[12]

Still, how do these *quasi-faculties* arise?

Every perception, idea, emotion, volition that appears in consciousness leaves a trace in the brain. These traces are not memories, per se, or they are not *just* memories. The idea of a trace is more inclusive than that of memory: an abstract idea, such as my idea of *dog*, is a collection of traces. The forging of temporary or permanent alliances between these abstract ideas constitutes reasoning. A syllogism is but the rapid succession of such ideas:

> All the impressions transmitted through the spinal cord, and those proceeding from the nerves of special sense, sight, hearing, etc., are collected in the brain. There, without being confounded the one with the other, they meet, are combined, associated, brought into the most manifold relations and combinations, and awaken within the brain other new, but purely subjective, internal images. All these images leave behind traces or remains, of which the combination produces again certain general results (Abstractions), and, quite involuntary, in the moment even of their production, they are already logically elaborated, collected, and associated in judgments, conclusions, etc.[13]

"Will" emerges when some of these traces latch onto the motor impulses in such a way as to direct their flow: "If, however, the known and definite ideas, by being united to the impulses of movement, exercise an influence upon the muscular movements, this is called Will."[14] I have the urge to eat, and I have the idea of Ratskellar München, and that urge fuses with that idea in such a way as to direct my steps to Ratskellar München: this is will. In the ordinary course of events, the strongest ideas, the most forcible and lively ones, pass over into action.

This brings us to a pivotal doctrine: the *motility* of the idea. For Griesinger, ideas are not stationary. They are not there to be contemplated or entertained.

They constitute a disturbance of the nervous system's preferred state, which is one of quiescence. The nervous system has the goal of attaining or maintaining this quiescence. *The "goal" of an idea, then, is to be discharged into action*:

> The ideas transform themselves into effort and will under the impulse of an internal force, in which we recognise, even in the innermost sphere of the life of the soul, the fundamental law of reflex action. We *must* will. In the healthy mind it urges and impels the individual to express his ideas, to realise them in actions, and thereby to rid himself of them. If this has taken place, the soul feels disburdened and freed; by the act it relieved itself of the ideas, and thus its equilibrium is again established. This is a remarkable fundamental fact of mental life which the inward experience of each man must know.[15]

Here, we see a dynamic portrait of the idea, a portrait that will play such a dominant role in Freud's writing.

But if ideas, by their very nature, tend to pass into action, then we have a puzzle: how do we prevent every fleeting fancy from transforming itself into action? I have the idea that I could remove a police officer's gun from its holster and shoot him with it. What prevents the idea from passing over, "automatically" as it were, into action? The reason it is not realized is that *the idea conflicts with other ideas which are stronger*, such as my idea of avoiding a life sentence in prison. Or, the idea of reaching for a bottle and the idea of maintaining my sobriety clash; if the idea of maintaining my sobriety is stronger than my idea of drinking, I will refrain.

> On this account the strongest ideas, at the end of their transition, pass forcibly into actions. It is a fortunate provision, however, in mental life that every perception does not attain to this degree of strength. Then, according to the laws of the association of ideas, there arise the contrasting perceptions; they draw after them further perceptions related to them, and there arises in consciousness a conflict.[16]

This is where the self enters the picture. *The function of the I is to mediate this clash of ideas.* Some ideas that enter consciousness harmonize with the *Ich*; other ideas clash with it. Those ideas which harmonize with the *Ich* are the ones that tend to be discharged into action: "The whole mass of ideas which represents the *I* comes into exercise, and gives the final decision according as it impedes or favors that first idea. This opposition in consciousness, which in the end is decided by the *I*, is the fact of the liberty of the individual."[17]

But whence the *I*? Whence this extraordinary presence and power? The *I* is but a massive cluster of traces that resonate with one another. It is a bundle of harmonious traces. We might liken it to the floating pool of garbage in the Pacific Ocean; it is just a conglomerate of ideas that are thematically similar and that therefore "stick":

> In the course of our lives, in consequence of the progressive combination of the perceptions, there are formed great masses of ideas which constantly become more associated. . . . The *I* is an abstraction in which traces of all former separate sensations, thoughts, and desires are contained, as it were, bundled together, and which, in the progress of the mental processes, supplies itself with new material; but this assimilation of the new ideas with the pre-existing *I* does not happen at once—it grows and strengthens very gradually, and that which is not yet assimilated appears as an opposition to the *I* as a *thou* [*ein Du*].[18]

Now, when certain ideas arise that do not fit in with the self-cluster, they are experienced as alien, as somehow external, as a *Du*. And the process by which the *I* forms is very fraught. As we grow, many different mid-sized clusters of ideas emerge, vying for dominance, vying for *I* status: for example, a mid-sized cluster associated with faith and a mid-sized cluster associated with disbelief vie for dominance, vie for incorporation into the *I* cluster. (We are reminded of Dennett's "bundle of semi-autonomous agencies.")[19] Self-education is about acquiring and solidifying one of these fundamental directions (*Grundrichtungen*).[20] With the emergence of the *Ich* and the *Du*, the stage is set for the production of madness, or at least one of the most paradigmatic forms thereof: the perversions of the intellect rather than those of sensation or will. As a result of the disease process—poor circulation, lesions, obstructions—certain ideas arise in the mind that are, in the technical sense, insane. These insane ideas are the ones that are not appropriate to the world, not properly calibrated to the world, not under the sway of the world. They represent the ideational byproduct of a nervous system that is not properly in tune with the world, that acts autonomously in relation to the world. When these insane ideas confront a healthy *I*, the *I* dismisses them, brushes them off like a bad dream. But as the disease progresses, the insane ideas gain a force and momentum of their own and they start to penetrate the mass of ideas that constitute the *I*; they begin to interleave themselves into this mass, they find openings, gaps, fissures, in which to insert their tendrils, one after the other; and when this process has reached its apex, when these insane ideas have perfectly infiltrated the *I*, the patient acquires a *false I*:

> In [insanity] also there is usually developed, with the commencement of the cerebral disease, masses of new sensations, instincts, and perceptions, proceeding from within outwards.... At first these stand opposed to the old *I* in the character of a foreign *thou* often exciting amazement and fear.... If the immediate cause of the new and abnormal state of the perception—the cerebral affection—be not removed, it becomes fixed and persistent; and because connections are gradually formed throughout with the groups of perceptions of the old *I*, and since frequently other masses of perceptions more capable of resistance are completely destroyed and effaced through the cerebral disease, the resistance of the old *I*, the struggle in consciousness, ceases by degrees and the emotions are allayed. But now, through these connections, through that introduction of abnormal elements of perception and will, the *I* itself is falsified, and has quite changed its nature.[21]

Diagnostically, the patient has become a different person. He was once religious and now he is a gambler; he was once a teetotaler and now he is a drunk. And so, this basic picture is of utmost importance for forensic purposes; the question of *was this person insane, or not, at the time of the crime?* must be answered by seeking biographical evidence of this falsification or alienation of the *I* from itself. And this basic picture is what therapy is all about: a goal of therapy, perhaps its most fundamental goal, is to intervene in this process as soon as possible before the insane ideas have managed to gain a foothold in the self-cluster. We must sever the tendrils before they insinuate themselves into the *I*.

Now we are in a position to confront the basic fact of madness. We are in a position to speak the apparently unspeakable. *At least in some cases*, madness would appear to happen not because the *I* has been forcibly hijacked by the insane ideas and passively dragged along by them. Madness happens when the *I* *acquiesces* to the insane ideas, when it lets them in. Recall that for Wigan, this was the almost unthinkable perversion of madness: that madness does not always overpower reason by force, but that reason, in its capacity as reason, chooses madness. Reason, in its capacity as reason, chooses madness, while knowing madness to be madness. This is not always the case. This is not a universal law for Griesinger. Sometimes madness *does* compel one, overpower one, despite one's valiant attempt to resist. Still, it is on the basis of this insight that we can understand why Griesinger thinks, *as a rule*, that delusions have the character of *wish fulfillments, just as do dreams*. Madness happens when reason knowingly and voluntarily abdicates its natural right of rulership: "partly she dragged him down; partly he sank."

We will start with Griesinger's characterization of dreams, because it is the ground upon which he builds his account of delusion:

> Agreeable, ravishing, heavenly dreams are very rare in health: they are most frequent in states of deep bodily or mental exhaustion, and we often observe at such times that the ideas suppressed during waking come forth strongly in dreams. To the individual who is distressed by bodily and mental troubles, *the dream realizes what reality has refused*—happiness and fortune. The starving Trenck, during his imprisonment, often dreamed of rich repasts; the beggar dreams that he is wealthy, the person who has lost by death some dear friend fondly dreams of the most intimate and lasting reunion.[22]

In dreams, one enjoys the satisfactions that have been denied during the day. This is, in fact, precisely the passage of Griesinger that Freud cites with great approval in *The Interpretation of Dreams* as a harbinger of his own theory of wish fulfillment.[23]

The same principle, however, is true of insanity:

> So also in mental disease, from the dark background of morbid painful emotion, by sinking into a still deeper state of dreaming, the repressed contending ideas and sentiments—bright ideas of fortune, greatness, eminence, riches, etc.; stand out—and, as soon as this happens, through a change in the state of the brain, but without recovery, the pressure of the painful sensations is removed, the former mental misery changes voluntarily to the mirth of the maniac. Thus we see clearly how supposed possession and imaginary realisation of good things and wishes, the denial or destruction of which furnished a moral cause of the disease, constitute commonly the chief subjects of the delirium of insanity: for example, she who has lost a darling child raves of a mother's joy, he who has suffered loss

of fortune imagines himself rich, the disappointed maiden is happy in the thought that she is tenderly loved by a faithful lover.[24]

How can we account for the fact that joyful delusions, for example, erotic delusions or delusions of grandeur, are more readily "fixed" than paranoid delusions? Because *we are more eager to welcome* those insane ideas that inflate the *Ich*, that confer extraordinary ability, extraordinary insight, extraordinary sexual prowess, than we are those ideas that belittle or terrify the self. Hence, in Griesinger, we rediscover the teleology of madness that we thought had been lost for good. Though madness always and necessarily results from brain pathology, at the same time, we have rediscovered its secret purpose, its secret utility, its secret function: often enough, reason chooses madness to buffer itself against the pain of living. Madness is a refuge and castle in an otherwise cruel world.

III.

The era of German imperial psychiatry, which we define as the confluence of these two ideas, the biologization of madness and the turn to a dysfunction-centered view of madness, culminates in the work of Emil Kraepelin, in two senses: in the sense of a particularly mature formulation of the doctrine, and also in the sense of a closure, for the era of psychoanalysis was about to begin. In fact, the massive sixth edition of Kraepelin's textbook coincided with the publication of Freud's *The Interpretation of Dreams* in 1899. Kraepelin eradicated even the small pockets, the tiny islands, of teleology that appear in Griesinger. Griesinger's teleology represented a strand of his thinking that ran counter to this double movement, the biologization of madness and the dysfunction-centered view of madness. His teleology only arises in his text to the extent that he is inappropriately relying on purely psychological concepts, *Ich* and *Du*: there are thoughts, emotions, and impulses that constitute the *Ich*, there are thoughts, emotions, and impulses that constitute the *Du*, and there is a drama—is the *Ich* going to resist these alien thoughts and emotions or is it going to acquiesce to them?

Such constructs are foreign to a thoroughgoing dysteleological approach to madness. And this is the sense in which Kraepelin's vision of a biological psychiatry was more radical than Griesinger's. It thoroughly expunged reference to teleology. It might seem strange to say that Kraepelin's biological vision of madness was *more* radical than Griesinger's, that he carried Griesinger's vision to its logical culmination. At least on the surface, Kraepelin was *resisting* the encroachment of biology into madness as much as he was advancing it. Kraepelin's mentor was Wundt, not Griesinger, nor the generation of psychiatrists that, following Griesinger, enthusiastically sought to discover the source of madness in the deranged tissue of the cortex. Kraepelin's preferred instruments were the rheostat and the chronometer, the favored tools of Wundt's experimental psychology, not the microscope. Kraepelin was of the opinion that the direct study of brain tissue in sane and mad populations would not be as revealing as precise, careful observation in the clinic and asylum. In the 1880s, he was preoccupied, like Freud, with Charcot's remarkable demonstrations with hypnosis and in 1888 he became proficient in hypnosis himself. In his posthumously published memoirs, he tells us that he was terrified that he would never have an academic career because he was not fixated on neuropathology in the way that most of his contemporaries were—the intellectual descendants of Griesinger, such as Wernicke, Hitzig, Fritsch, Flechsig, and Meynert. In fact, the reason that he began writing his textbook in 1883 was not because it was a passion, but because he needed the money; such was his struggle to find gainful employment.[25] Even today, when we think of the "Kraepelinian legacy," our first thought is *not* the study of neuropathology but careful, empirically oriented diagnosis and classification, with an emphasis on longitudinal course.

Still, we must insist that Kraepelin radicalized Griesinger's biological-dysfunction-centered legacy by fusing that legacy, irrevocably, with the project of classification. With Kraepelin, these were no longer *two* quite independent projects: the project of the neuropathologist seeking the organic source of madness, and the project of the nosologist attempting to come up with a convenient system of classification. The very purpose of classification, for Kraepelin, is to contribute to the project of *identifying and exposing this elusive disease entity*. There are different ways of "identifying" an organic dysfunction underlying madness. The first is to dissect the mad person's brain and to examine the tissue slices under a microscope. The second is to use rigorous empirical observation to pin down the developmental sequence of the disease, its intrinsic temporality, its progression. In other words, by identifying the chronological sequence of the disease, you are able to pose, precisely, the follow question: *what kind of organic process must this disease be such that it exhibits this particular chronological dynamic?* In the jargon of the philosopher of science, we can say that for Kraepelin, disease categories are *mechanism-sketches.*[26] When I say, "Your husband has dementia praecox, he is not paranoid," what I am saying is that the neuropathic entity that generates his type of speaking and feeling and acting is different from the neuropathic entity that generates paranoid delusions, and here is its course: if we do not intervene early it will lead to a progressive mental deterioration; he will quite literally become a "back ward" case. (Incidentally, Kraepelin, early in his career, worked in the district mental asylum in Munich; he worked in ward G, which was quite literally a ward *behind* the other wards, a *back ward*, the ward in which only the most hopeless and violent patients were housed.)[27] Hence the urgency of psychiatric classification: it is not something that comes *after* we discover the inner mechanism of madness; classification is not independent of the task of discovering the mechanisms of madness; it is a propaedeutic to discovering those mechanisms:

> I was forced to realize that in a frighteningly large number of patients, who at first seemed to have the syndrome of a mania, melancholia, insanity, amentia, or madness, the syndrome changed fairly quickly into a typical progressive dementia. . . . I soon realized that the abnormalities at the beginning of the disease had no decisive importance compared to the course of the illness leading to the particular final state of the disease. . . . Bearing this in mind, it seemed particularly important to recognize the disease in good time and to differentiate it from other similar syndromes.[28]

Does that mean that there is no teleology in Kraepelin's account of madness? Teleology exists, *but only in the form of the conscious idea.* To the extent that it is remotely plausible to look at a certain symptom as goal-oriented, as purposeful, as functional, as adaptive, it is because we are seeing it as the result of more or less conscious intentionality: teleology, for Kraepelin, at least before the psychoanalytic revolution, is the hallmark, the sure sign, of malingering.

There are at least two places in Kraepelin's corpus where he deals squarely with the apparent purposiveness or goal-directedness of certain symptoms. The first and earlier of these is in his discussion of acquired neurasthenia in the seventh (1903) edition of the *Lehrbuch*. Here, the chronic invalidity of the neurasthenic patient represents a *strategy* for perpetuating the sick role and acquiring its associated benefits. In the most extreme cases, he tells us,

> the patients tend to become chronic invalids of a most distressing type. . . .
> They betake themselves to the seclusion of a charitable institution with its

freedom from annoyances, or if they remain at home, demand the utmost consideration for every whim. They have no thought for the maintenance of the family or appreciation of the burden which they create. The increasing demand for sympathy leads to prevarications and to various assumed contortions, in order to assure the physicians or friends that they are in critical condition.[29]

Acquired neurasthenia is just artful malingering. *Nor could it be otherwise*: for to the extent that madness has a purpose, to the extent that it is *by design*, from where, exactly, would it draw its purposeful character? Whence its design, *if not from the purposiveness of the conscious idea, of conscious intentionality*? And hence, how could it *not* be a form of malingering? It certainly is not drawing its purposeful character from the will of God, as in Burton. Kraepelinian psychiatry has no room for a mysterious vital principle, the *vis conservatrix* of Pinel. By now we have fully entered Taylor's secular age.[30] And we have yet to enter the era in which psychoanalysis, and the idea of the unconscious wish, has become commonplace. The lofty talk that we find in Griesinger and Wigan of the confrontation between self and other, this talk of the self either valiantly resisting the encroachments of madness or succumbing to them, does not exist in Kraepelin's system; there is no room for such psychological constructs there. This does not mean that Kraepelin does not believe in a self, in an *I*; it is just that the *I* does not play any deep explanatory role in making sense of madness. It is not that Kraepelin simply neglected or ignored teleology. It is not that he was so entranced with "medicalization," whatever that turns out to be, that it never occurred to him, prior to his exposure to psychoanalysis, that madness has a secret purpose. Madness and teleology are logically exclusive. They cancel each other out.

One further confrontation that Kraepelin has with the apparent goal-directedness of some disorders occurs much later in his life, during World War I, with the war neuroses. Here is the basic narrative: during the war, there are reports of people coming back from the battlefield with peculiar symptoms; in some parts of the world this is referred to as "shell shock," no doubt to avoid the implication that these symptoms are psychogenic in nature. "Shell shock" carries the implication that these symptoms are literally repercussions of having a bomb explode near you; the proximity of the bomb has induced a pathology of the nervous system. But it did not take long for the psychogenic nature of the phenomenon to become obvious to many researchers: people would experience "shell shock" who had not been very close to combat, and, more tellingly, the symptoms would dissipate as soon as the soldiers were sent back home. It became impossible not to see in the phenomenon a kind of purposiveness, a goal-directedness.[31]

By World War I, Freud and his followers had a rich psychological framework within which they could pose the problem of shell shock. They were finally in a position to answer Kraepelin's question of the source of teleology: it originates not from the teleology of the conscious idea, but the unconscious idea. They could explain the phenomenon in the same terms with which they understood hysteria. It is a form of conversion hysteria. There is no question of malingering or duplicity. This is an unconscious mechanism for avoiding a traumatic memory; the patient is avoiding confrontation with those memories through their somatization, by displacing them onto the body. These are defense mechanisms that operate outside of the patient's conscious awareness or control.[32] But absent that kind of rich psychodynamic framework, it is difficult to see in the war neuroses anything but simple malingering.

At least at the outset of the war, this was exactly Kraepelin's point of view, as we discover through his memoirs. It is difficult to read his accounts of veterans suffering from battle fatigue, at least at the outset of the war, as anything other than sheer mockery:

> Due to the length of the war more inferior persons had to be recruited and battle fatigue increased. Therefore, all kinds of more or less distinct symptoms meant not only a long-term stay in a military hospital, but also dismissal from the army and an appropriate pension, which had disastrous effects. Public sympathy was aroused by the apparently badly injured "Kriegszitterer" [literally "war-shakers"], who attracted general attention on the streets and were showered with gifts. Under such circumstances, the number of those, who believed themselves to be entitled to a dismissal and further support, because they had shock or had been buried alive, increased to flood-like proportions.[33]

It is true that as the war continued, and as the tools of psychoanalysis began to help soldiers get back on the battlefield, Kraepelin began to soften his judgment. He tells us that, in 1917, he was brought to a field clinic where he saw not only the effectiveness of hypnosis, but also certain "intense mental influences" whose nature he chose to leave unspecified, in leading combat veterans to an adequate recovery—or at least adequate enough to resume the fight.[34] For Kraepelin, the symptoms were no longer *simple* malingering, but had a deeper psychological source: "What we could observe amongst patients suffering from rheumatism and sciatica also applied to the patients with heart, stomach and lung diseases; hidden behind the various masks was the need to escape from the terrible pressure of the war."[35] Still, he saw war neurosis as the sign of a fundamentally weak and morally dubious character, even after his encounters with hypnosis and other, unspecified, psychological treatments. He expressed profound disappointment over having to replace injured soldiers in his clinic with those suffering battle fatigue; invariably, the latter were people of flimsy moral character: "It was evident at first glimpse that we were dealing almost exclusively with inferior, incapable and frequently even malicious personalities."[36] While war neurosis is no longer malingering, it is surely a sign of weakness that one would even be susceptible to conversion, somatization, etc.

In this context, we can begin to appreciate the striking originality of Freud's vision of the relationship between madness and teleology. Freud gave us a fundamentally new way of understanding the teleology of madness. Madness, to the extent that it can be seen as purposive, does not draw its purposive character from the will of God, or from a mysterious vital principle, or from the teleology of conscious intentionality. It draws its purposive character from an *unconscious* wish and an equally *unconscious* impetus to prevent that wish from becoming conscious. All madness can be traced to the interplay of these equally purposive, goal-directed, but non-conscious agencies.

PART III

MADNESS AND THE GOAL OF EVOLUTION

In the twentieth century, madness-as-dysfunction and madness-as-strategy sit in awkward proximity to each other. *On the one hand*, the century is capped, on both ends, with a thoroughgoing teleological orientation: psychoanalysis at the beginning of the century and evolutionary psychology at the end. The century opens with psychoanalysis, with its thesis that nearly all forms of psychopathology are strategies for preventing forbidden wishes from gaining access to consciousness, while partly fulfilling them. At the end of the century, we have Darwinian medicine, which depicts many forms of mental illness, such as depression, the anxiety disorders, and even symptoms of schizophrenia, as adaptations, as having evolved purposes, as having been designed by natural selection to cope with Pleistocene existence. *On the other hand*, it is equally just to call the twentieth century an era of dysteleology. We are repeatedly told that mental disorders have been, or are on their way to being, completely explained in terms of neurological, developmental, or genetic dysfunctions. At the same time, our classification systems, most notably the *Diagnostic and Statistical Manual of Mental Disorders* (*DSM*), as well as the Research Domain Criteria project (RDoC), assure us that *a classification of mental disorders is just a catalogue of all of the ways the mind can err*: the *DSM-III* tells us that mental disorders necessarily spring from "behavioral, psychological, or biological dysfunction[s]," and the founders of the RDoC project tell us, in the same vein, that "mental disorders can be addressed as disorders of brain circuits."[1]

Perhaps what is most extraordinary about psychiatry in the early twenty-first century is that almost nobody seems to have noticed the tension, even contradiction, between madness-as-dysfunction and the evolutionary approach to psychiatry: the extent to which one is willing to adopt an evolutionary approach is precisely the extent to which one is willing to entertain the hypothesis that some mental disorders are adaptations, not dysfunctions; strategies, not failures. The point that we must repeatedly insist upon here is that *we need not take up one stance or the other*; our goal is not to establish that mental disorders "really are" biological dysfunctions or that they "really are" adaptations or strategies. The point is to render the tension, even the contradiction, in contemporary psychiatric discourse *manifest*, and thereby carve out a new intellectual space for thinking about what madness might be.

Freud opens the century by placing psychiatry squarely within a teleological context: all forms of mental illness are nothing but strategies for satisfying forbidden wishes (they are "wish fulfillment strategies"). Perhaps one of the most powerful and enduring legacies of psychoanalysis was the extent to which this basic teleological orientation was embedded into classification: the *DSM-I* of 1952 is constructed entirely around this insight. In that manual, the three basic forms of non-organic mental illness—the psychotic, neurotic, and personality disorders—are explicitly presented as *three strategies that the mind uses for coping with stressors*. The project of future editions of the *DSM*, from the *DSM-II* onward, is to expunge, systematically, reference to teleology.

Kurt Goldstein takes this psychoanalytic picture and *biologizes* it, much like Griesinger biologized Kant's faculty psychology. He thinks of the teleology of madness in terms of the progressive ontogenetic adaptation of the organism to its environment, and, reciprocally, the construction of the environment by the organism. It is true that many forms of mental illness stem from brain injury—to that extent, Goldstein is committed to a dysfunction-centered model. Still, we must see, at the same time, not only in mental illness but in disease generally, the working out of what he calls the *self-actualization of the organism*. Disease does not represent a stunting of this project of self-actualization; it represents the *funneling* of this project of self-actualization though a narrower channel.

In the early part of the century, psychoanalysts like Frieda Fromm-Reichmann and Harry Stack Sullivan showed us how the forms of schizophrenia could be seen as coping mechanisms, strategic retreats from a painful reality. With the double-bind theory associated with Bateson and later developed by R. D. Laing, madness is no longer thought of *merely* as a passive withdrawal from the social world, but as an active form of participation within that world. It is ultimately a political act: an assertive, steadfast *refusal to participate* in a corrupt social order. The mad person is fused with the revolutionary. Madness is an instrument for defeating fascism.

In the late 1960s, as noted above, starting with the *DSM-II*, continuing with the *DSM-III* and its revisions, and culminating with the RDoC, we witness the progressive attempt to eliminate reference to teleology in mental disorder classification, just as Kraepelin attempted to do. For Kraepelin, there was simply *no room* for teleology in madness: to the extent that we see purposiveness and goal-directedness in madness, we are dealing with malingering rather than madness. Similarly, the *DSM-II* should be seen as a sustained attempt to complete the process that the *DSM-I* initiated but failed to carry out: purging teleology from nosology.

By the 1990s, we had become fully immersed in the era of Darwinian medicine, which is just as likely to see any given mental disorder, such as depression, as an adaptation to a Pleistocene environment as it is to see it as a brain dysfunction. What is astonishing about this is simply the fact that, first, almost nobody has seen Darwinian medicine as merely the culmination, the latest episode, in a centuries-old style of psychiatric thought, and, second, almost nobody has appreciated the deep tension between the evolutionary approach and the biological dysfunction approach. A major purpose of this entire book is simply to reorganize our thinking in such a way that this tension becomes obvious.

11

The Strategies of Wish Fulfillment

In 1899, with *The Interpretation of Dreams*, Freud advanced a teleological approach to virtually *all* forms of madness: the forms of madness are just so many different strategies for preventing a forbidden idea from becoming conscious, while arranging for its deviant fulfillment. In fact, he warns us repeatedly *not to mistake a functional performance for a dysfunctional one* (Section I). This basic picture of madness is rooted in a view about the nature of the nervous system he shares with Griesinger: the goal of the nervous system is to maintain quiescence; mental activity (ideas, wishes, emotions) disrupts that quiescence, and therefore it must be discharged. Unlike Griesinger, however, Freud understands that some ideas are not merely *alien*, that is, they contradict one's self-conception, but are *forbidden*, that is, they cannot become conscious without destabilizing the self (Section II). Freud's principle for understanding madness, that the forms of madness are strategies for partially fulfilling forbidden wishes while preventing them from becoming conscious, yields a principle of classification, too. Mental disorders can be distinguished in terms of the precise nature of the strategy that the mind deploys for this end. One of the most enduring legacies of Freud was the way the first edition of the *Diagnostic and Statistical Manual of Mental Disorders* of 1952 builds upon his teleological approach to classification.

I.

The Interpretation of Dreams announces three new principles for rethinking the relationship between madness and teleology. The first is that *all forms of madness are functional*. They have a hidden purpose, a secret end. The goal of the healer is to make this purpose known. But whence does madness draw this secret purpose? For Burton, the purpose of madness had its source in the purposes of God; for Pinel, its purposiveness was derived from a mysterious vital principle. Kraepelin could only see in madness, to the extent that he observed goal-directedness at all, the purposiveness of the conscious

Madness. Justin Garson, Oxford University Press. © Oxford University Press 2022.
DOI: 10.1093/oso/9780197613832.003.0012

idea: malingering. Freud introduced a new, hitherto unthought source of teleology: the teleology of the unconscious wish. And this is his second principle: *the purposiveness of madness is derived from the purposiveness of a wish
that must not be allowed into the privileged circle of consciousness.*

To be more precise, the purposiveness of madness results from the interplay of two psychological agents (currents, systems, forces, etc.), each of
which has its own function or job. The first is that of the unconscious wish,
which strives for realization—or better, the psychological system that strives
to bring that wish to fulfillment, that is, to discharge it, to release the heightened excitation, the cathexis (*Besetzung*) that it represents. The second is the
purposiveness or the function of a censor/repressor, which has a two-fold
role. It categorically prevents the wish from migrating into the privileged
circle of consciousness, in such a way, however, as to fulfill that wish, at least
in part. The convergence of these two teleological forces yields the *psychopathological structure.*

His third insight, and one of equal importance, is that *the forms of madness
are different strategies that the psychic system utilizes for discharging (that is,
satisfying) the forbidden wish*, at least partially. That is to say, these forms are
so many different strategies for keeping the wish outside of the privileged
circle of consciousness while allowing its partial fulfillment. This implies
a new system of psychiatric classification, an *anti-Kraepelinian*, even *anti-
Kantian*, system: each form of madness is a different strategy for fulfilling/
discharging the forbidden wish.

Freud states repeatedly throughout his entire corpus that, *as a rule*, we
must *not* attempt to trace madness to some underlying dysfunction, pathology, disease: the forms of madness, in their deepest sense, are not
dysfunctions; they serve functions; they have purposes; each has its raison
d'être. The mad mind is a well-functioning machine whose goal and mode of
operation remain to be discovered. This is not to deny that mental disorders
are ultimately biological in nature. They can and should ultimately be traced
to neurological entities.[1] But when that day arrives, when we can restate
the entirety of the Freudian apparatus in neurological terms, what we will
find is that the various forms of madness represent the proper functioning
of the nervous system, not its pathological dysfunction. That is, on that day
where we can trace all madness to the inner workings of the nervous system,
the core function of which—as we know from his posthumously published
Project for a Scientific Psychology—is to minimize unnecessary levels of excitation while preserving the integrity of consciousness, we will discover that

the forms of madness, its distinctive systems, which he calls psychopatho-logical structures (*eine psychopathologische Bildung*) are circuitous routes for achieving that goal; in madness the nervous system is working exactly as it should and as it must.[2]

Freud constantly reminds his readers that the products of what we think of as inner dysfunctions are, in fact, the outcomes of *the proper functioning of the psychic apparatus*: they are strategies for accomplishing goals. Consider his 1911 paper on Judge Schreber, "Psycho-analytic Notes on an Autobiographical Account of a Case of Paranoia (Dementia Paranoides)":

> And the paranoiac builds [his world] up again, not more splendid, it is true, but at least so that he can once more live in it. He builds it up by the work of his delusions. *The delusional formation, which we take to be the pathological product, is in reality an attempt at recovery, a process of reconstruction.*[3]

Or his 1901 *The Psychopathology of Everyday Life*:

> The mechanism of parapraxes and chance actions . . . can be seen to cor-respond in its most essential points with the mechanism of dream-forma-tion. . . . In both cases *the appearance of an incorrect function is explained by the peculiar mutual interference between two or several correct functions.*[4]

Finally, consider his short essay, "On the Sexual Theories of Children," of 1908:

> The only difference [between healthy people and neurotics] is that healthy people know how to overcome those complexes without any gross damage demonstrable in practical life, whereas in nervous cases the suppression of the complexes succeeds only at the price of costly substitutive formations— that is to say, from a practical point of view it is a failure.[5]

Here, the neurosis is a *failure*, to be sure, but not in the sense that anything has, as it were, *erred in its function*. Rather it is a failure in that the strategy that the patient has deployed, the basic gambit that the patient has made, has, in practical terms, fallen short of its goal. It is a kind of failure that presupposes that *the neurotic patient is trying to do something with her neurosis*; she is up to something; the healer's job is to find out what she is up to.

II.

The purposive character of madness is grounded, is implemented and realized, by the functionality of the nervous system. We must, therefore, begin with a basic vision of what the nervous system is "trying to do," of what it is all about. The shallow answer is that its ultimate purpose is drawn from the purpose of the organism, which is to perpetuate the life of the species. But *all* parts of the organism, at that level of analysis, have exactly the same function: to help us survive and breed. What, then, is the nervous system's *specific* function? What is its signal contribution to the larger project of surviving and breeding? Its specific function, one might say—a little less shallowly—is to gather information about its environment and use that information to select an optimal course of action. The purpose of the brain, as Griesinger had already indicated in 1845, is to compile something like (as we would now say) a library of "sensorimotor contingencies."

But even this apparently sophisticated way of describing the function of the nervous system is still shallow, because it still fails to state its purpose in such a manner as to render its operation *perspicuous*. It is like saying, "The purpose of the nectar glands in flowers is cross-fertilization." Of course it is, but that does not state its function in a manner that makes its ordinary operations immediately explicable. It does not render its operations perspicuous enough to allow us to answer very basic questions about it, like: *why is the nectar gland located deep within the flower rather than on its surface?* Once we understand, however, that its *specific* function is to attract insects, and *thereby* get the insects to carry its pollen, and *thereby* assist in the process of cross-fertilization, then we understand immediately why the nectar gland *must* be located deep within the flower: because the pollen must attach to the insect's body and one way to ensure that is to force the insect to crawl deep within the flower to obtain nectar. We also understand why it is that only brilliantly colored flowers have nectar glands, why the activity of the nectar glands is synchronized with the circadian rhythm of the favored insect, and so on. So, while all of these function attributions are technically *true* ("the function of the nectar gland is to perpetuate the species," "the function of the nectar gland is cross-fertilization," "the function of the nectar gland is to attract insects"), they can be stated in a way that is more or less revealing.[6]

The function of the nervous system, what it is designed to do, that is, its *most specific function*, as Freud tells us in his posthumously published *Project*, is to achieve and maintain a state of quiescence.[7] It is like Minsky's "useless machine" whose only function is to turn itself off. Every time the sensory organs are irritated—light, pressure, sound—the nervous system enters an increased state of agitation, a filling or *Besetzung* (which Strachey translates as "cathexis"), a heightened state of excitation that the nervous system attempts to discharge. Its psychological correlate is unpleasure. And the psychological tendency toward avoiding unpleasure is, at root, one with the nervous system's drive to discharge energy: "Since we have certain knowledge of a trend in psychical life towards *avoiding unpleasure*, we are tempted to identify that trend with the primary trend toward inertia."[8] Technically speaking, the goal of the nervous system is not to maintain *absolute* quiescence, which would make animal life impossible, but to maintain a *non-zero but low* degree of activation.[9] It must retain a low degree of activation because it needs a reserve in order to respond appropriately to unexpected stimuli.

And that *Besetzung*, that heightened excitation or filling, is typically discharged through the motor system, through movement. As we grow from infancy onward, we learn which motor sequences lead most efficiently to the discharge of heightened activation. An infant learns early on that the cathexis associated with the sensation of hunger is most effectively discharged by wailing; later she learns that this cathexis is most effectively discharged by putting food in her mouth. This way of seeing the function of the nervous system renders its ordinary operation explicable in a way that all other formulations fail to. For example, Freud was an early proponent of the synapse doctrine, and he was able to deduce the reality of the synapse from his grasp of the function of the nervous system: he understood that *if* the nervous system's goal is simply to discharge excess energy, and *if* this cathexis is somehow or other associated with a heightened activity of neurons, *then* there must be some kind of an obstruction or blockade—a *contact barrier*—that stops the neuron from immediately discharging its activation.[10] By analogy, for an electrical current to actually make anything work, to be productive, there must be resistances built into the system to reroute and divert the flow of electricity. In Freud's thinking of the function of the nervous system we see a deep resonance with Griesinger, who also describes thought as an irritation or a disturbance of the neural tissue, an irritation that must be discharged. The vital consequence for the mind is that the mind is composed of *motile* elements: every thought, every desire, every sensation, every wish, is alive and in motion; it seeks expression, discharge, liberation. Ideas are not stationary; they are not there to be contemplated. They are there to propel the organism along until they extinguish themselves in action. This is true of all

mental elements—thoughts, ideas, wishes: all of them are but disturbances of
the neural tissue and demand to be discharged.

This understanding of the "psychic apparatus," however, leads to a
puzzle: what happens when a wish cannot be fulfilled, because it contradicts
the demands of reality, or because it contradicts the character of the self, or
because it is socially unacceptable? It cannot just be "bottled up"; it does
not evaporate; that is not how ideas work. The wish must seek a *deviant*
mode of fulfillment. In retrospect, it should not seem too surprising that
Griesinger maintained that dreams are wish fulfillments half a century before Freud. Like Freud, Griesinger even expanded this to see clearly that
some psychotic systems of delusion, like dreams, are also wish fulfillments.
For Griesinger, as for Freud, it is necessary that at least some dreams and
some delusions are wish fulfillments, because the whole structure of the
nervous system is designed in such a way that all wishes must somehow be
discharged. If they cannot be satisfied during the day, while we are awake,
then they must be satisfied at night, while we sleep. Hence Freud's only reference to Griesinger in *The Interpretation of Dreams* is overwhelmingly approving: he writes of Griesinger's "acute observation" that "ideas in dreams
and in psychoses have in common the characteristic of being *fulfillments of
wishes*."[11]

We said that some ideas cannot be fulfilled, not because they are socially
unacceptable, but because they contradict the character of the self. An idea
that arises inside of me—say, to strike someone in the face—contradicts my
idea of myself as a peaceful person. This raises a general question: how, precisely, do these wayward ideas relate to this entity called the *self*, Griesinger's
Ich? Ideas originate through cortical disturbances. The *I* is not, as it were, the
author of these ideas. Many of our ideas are triggered mechanically or spon-
taneously, in a manner that is completely independent of the character of the
self. There is no a priori guarantee, therefore, that the ideas that arise inside
of me are consistent with my sense of self. For Griesinger, as we have seen,
the *I* is simply a massive cluster of ideas or traces with a common theme,
a common valence or orientation. Given that ideas emerge in a somewhat
spontaneous way, and that the *I* is not the author of them, *some* of the ideas
will be consistent with this conglomerate and will be assimilated, and *others*
will be inconsistent and will be discarded as foreign, as "other." ("Why did
I have that weird urge to strike that man in the face? That's not me at all.") The
ones considered foreign will be shunned; they will be prevented from seeing
the light of day. This was the basis for Griesinger's split between the *Ich* and

THE STRATEGIES OF WISH FULFILLMENT   185

Du. And these are precisely the ideas that must seek deviant modes of fulfillment, most notably through dreams.

For Griesinger, all of this is fairly uncomplicated, and that is because Griesinger failed to recognize something that Freud grasped very
clearly: there are ideas that are not *simply* alien, that do not *simply* make us
shudder, but *that are forbidden.* They cannot be brought before the *I* without
fundamentally destabilizing it. Some ideas, in other words, are not just alien
and to be shrugged off, but must be prohibited from entering the magic circle
of consciousness. But what gives them their destabilizing power? Why can
they not simply be shrugged off as alien, as perverse? If I am caring for an
elderly father, and I have the thought *I wish the old man would die already,*
I cannot shrug that off as an alien thought the way I would shrug off a piece
of music stuck in my mind: *Why is Gloria Gaynor's "I Will Survive" stuck in
my head?* The latter can be shrugged off as a random bit of ideation. The idea
Why won't the old man die already? so nakedly reveals an unmistakable desire
that it *simply cannot be thought*; it cannot be entertained. The dilemma is that
those forbidden ideas demand fulfillment just as all the others do. So, Freud
is struggling openly with a problem that was invisible to Griesinger: how to
discharge the *Besetzung* associated with a forbidden idea?

To solve this problem, we need a new mechanism, a very special kind of
machine, one that, like these forbidden wishes, operates in the subterranean
space of the mind, in the nether regions. This system is what he calls in various places the censor, repression, or the pre-conscious system. This agency
has two roles, a *dual teleology*: it *arrests* the movement of the unconscious
wish before it passes into consciousness, and it *arranges* for its deviant fulfillment. It prohibits the idea from entering consciousness while enabling its
partial fulfillment. *All* forms of madness as well as dreams and parapraxes
represent compromises between these two systems or forces. He sometimes
calls the products that issue from this interaction "compromise formations"
(*Kompromissbildungen*).[12] But more tellingly, as in *The Interpretation of
Dreams*, he calls them "psychopathological structures" (*psychopathologische
Bildung*).[13] The very expression "psychopathological structure" would seem
to be infelicitous: have we not insisted all along that these structures are
not pathological, that they are not diseases, that they do not represent the
product of an inner dysfunction? Surely, the more sanitized term "compromise formation" is preferable because it does not connote the idea of disease
or dysfunction. While that is a valid concern, the very idea of a *psychopathological structure* is valuable because of the way that it joins two apparently

incompatible ideas: the idea of pathology and the idea of structure, the idea of disorder and the idea of order. Surely, pathology is nothing if not disorder itself. Pathology is the breakdown of structure; it is the violation of order. Today we often talk about diseases as resulting from the *breakdown* of a mechanism for this or that; Alzheimer's results from the *breakdown* of a mechanism for memory; drug addiction from the *breakdown* of a mechanism for reinforcement learning; autoimmune disorders from the *breakdown* of the mechanism for self-recognition. *Mechanism* is the principle of order, organization, function; *breakdown* is the force or power that leads to its dissolution. "Psychopathological structure" carries, inside of itself, a kind of tension; it appears as an oxymoron, a contradiction in terms. The term "psychopathological structure" points to a very peculiar sort of artifact.

Consider compulsive handwashing. For Freud, this is a *psychopathological structure*, which serves a two-fold purpose: it is a means of *expiation* for a moral crime and, at the same time, it keeps the criminal idea outside the sphere of consciousness.[14] It is a brilliantly designed artifact with a dual function. We must, then, understand the term "structure" in "psychopathological structure" as correcting, or rebutting, a connotation buried in the word "psychopathological." The term "psychopathological structure" is not a contradiction in terms: rather, the second term reverses, or neutralizes, the connotation of the first. Consider "expectant mother" or "rising star": the adjective negates the proper application of the noun; it evokes a certain concept and, at the same time, denies the appropriateness of that concept.

Freud repeatedly demands an inversion of thought: what we think of as dysfunctional must be thought of as functional. One of the most remarkable examples of this tendency is in *The Psychopathology of Everyday Life*. Strachey uses the term *parapraxis* (e.g., slips of the tongue, misremembering, slips of the pen) to translate *Fehlleistung*, but it would be more accurately rendered as "faulty function." The term, then, would appear to point to some sort of *dysfunction*, or *malfunction*, or *breakdown of function*. But of course, for Freud, *every* Fehlleistung *serves a function*: on the very first page of the book, he tells us that forgetting—one of the parapraxes—serves a function: when we forget a name, it is not because memory fails to operate; rather, memory *refuses to operate*.[15]

Having appreciated the teleological character of Freud's vision, we can return to one of his earliest texts, his *Studies in Hysteria* with Breuer, and discern the tension that animates the text: Freud and Breuer brought two fundamentally conflicting perspectives to bear on the problem of hysteria.

Psychoanalysts Mitchell and Black, in their excellent overview of twentieth-century psychoanalytic thought, emphasize this divergence.[16] Breuer represents, and never departs from, *madness-as-dysfunction*—a patholog-ical orientation, a disease model of hysteria—while Freud, in the sections he wrote, introduces and insists upon *madness-as-strategy*. This is apparent in the very metaphors that Breuer repeatedly uses. His core theory of hysterical conversion is that when the cerebral excitation associated with the repressed experience surpasses a specific threshold, it destroys the normal barrier be-tween the psychic life and the motor system. Using an electrical metaphor, he likens this process to a "short circuit."[17] Conversion symptoms, for Breuer, result from the dysfunctional breakdown of a barrier between psyche and soma: the barrier cannot perform its function. Freud's perspective could not be more different. In the sections that he authors, he tells us that hysteria al-ways has a *motive* and a *mechanism*.[18] The *motive* is the desire to avoid the repressed idea, the traumatic experience. The *mechanism* is the conversion of the energy of the repressed idea, its cathexis, from the mental to the bodily sphere. Hysteria is not the result of a broken mechanism but the result of a mechanism that functions exactly as designed.

These conflicting metaphors feed into a broader tension between Freud and Breuer about the etiology of hysteria: while both Breuer and Freud recognize that hysterical conversions can result from the repression of a memory, and they can serve the function of preventing that memory from entering consciousness, Breuer downplays the significance of that teleolog-ical mechanism in the etiology of hysteria. While he tells us that all instances of hysteria come from an idea that cannot be assimilated to conscious life, he also tells us that there are two *different* reasons it cannot be assimilated. The *first* is defense (*Abwehr*): the experience is consciously repressed because the person does not want to confront it. Breuer attributes this picture to Freud.[19] The *second* is that it cannot be assimilated because that experience took place in a kind of altered state of consciousness and therefore it was never ade-quately integrated into conscious life in the first place. There is no question here of repression: it is because a certain idea was received in an altered state of consciousness that it never quite made it into the "mainstream" of con-scious life. And then he states in no uncertain terms that the latter is the more prominent mechanism: "The latter seem to be of the highest importance for the theory of hysteria, and accordingly deserve a somewhat fuller examina-tion."[20] It is not just that Breuer is more prone to use metaphors for broken machines—short circuits, breaking, and so on; it is that he does not think conversion symptoms, as a rule, have any special function or purpose at all. They are mere byproducts of the fact that the ideas were received in an altered state of consciousness.

III.

Freud demands that we adopt a radically different starting point for the project of identifying and enumerating the diverse forms madness can take: this must be a fundamentally *anti-Kantian* starting point, because, for Kant, the forms of madness are nothing but aberrations of function in the corresponding faculties of the mind. For Kraepelin, as for Kant, the whole point of distinguishing the various mental disorders, for example, distinguishing dementia praecox and paranoid psychosis, is to track, expose, and arrest the underlying disease processes. For Freud, the forms of mental illness are, instead, so many different strategies that the mind uses to fulfill its twofold function of keeping forbidden desires out of consciousness while orchestrating their deviant fulfillment. The different disorders, at least in their most general division, are just so many strategies for carrying out this task. Freud announces this basic framework in his 1894 "The Neuro-Psychoses of Defence," and this teleological framework was ultimately incorporated into the first edition of the *DSM*, in 1952. Psychiatric classification, in the second half of the twentieth century, represents a sustained attempt to eradicate these remnants of teleology.

At the beginning of his 1894 paper, Freud announces his ambitious intent to explain the diverse forms of madness—hysteria, phobias and obsessions, and hallucinatory psychoses—in terms of a common etiology. They all originate from a moment in which "an occurrence of incompatibility took place in [the patient's] ideational life," that is, from when an "experience, an idea, or a feeling" clashed with their conception of self. This clash produced such a distressing affect that "the subject decided to forget about it because he had no confidence in his power to resolve the contradiction between that incompatible idea and his ego by means of thought-activity."[21] All disorders *share a common goal*, namely, of "*turning this powerful idea into a weak one*, in robbing it of the affect—the sum of excitation—with which it is loaded."[22] Extraordinarily enough, this basic teleological orientation never changes in Freud, from 1894 until the end of his life, despite the extremely deep and thoroughgoing intellectual transformations that Freud undergoes during these years. This core insight, that the basic forms of madness are strategies for transforming a stronger idea into a weaker one, precedes his discovery of the Oedipus complex; it precedes infantile sexuality; it precedes the death drive. This is a time when he still earnestly believes that, often enough, the memory the patient is trying to repress is the memory of an actual instance of sexual abuse, a memory that has been reactivated as a result of the passage into puberty and that, consequently, takes on an additional filling, a *Besetzung*.

What are these basic strategies? Freud recognizes three: *conversion, transposition*, and *frank denial*. In hysteria, the forbidden idea is *converted* into a bodily symptom.[23] In the obsessions and phobias, the distressing idea is separated from its corresponding affect; the newly liberated affect is then *transposed* onto a different idea; it "*attaches itself to other ideas which are not in themselves incompatible; and, thanks to this 'false connection,' those ideas turn into obsessional ideas.*"[24] In the final case, the distressing idea or experience is resolved by a *frank denial* of its very existence. Unfortunately, in order to deny the existence of this idea, the patient is forced to deny large swaths of her own perceptual experience. For example, "the mother who has fallen ill from the loss of her baby, and now rocks a piece of wood unceasingly in her arms, or the jilted bride who, arrayed in her wedding-dress, has for years been waiting for her bridegroom," have both adopted the risky strategy of resolutely denying the existence of the painful event.[25]

This picture is elaborated in his "Further Remarks on the Neuro-Psychoses of Defence" of 1896, but here, he deems obsessional neurosis and paranoia to be two *different* strategies for resolving repressed thoughts of self-reproach:

> In obsessional neurosis the initial self-reproach has been repressed by the formation of the primary symptom of defence: *self-distrust*. With this, the self-reproach is acknowledged as justified; and, to weigh against this, the conscientiousness which the subject has acquired during his healthy interval now protects him from giving credence to the self-reproaches which return in the form of obsessional ideas. In paranoia, the self-reproach is repressed in a manner which may be described as *projection*. It is repressed by erecting the defensive symptom of *distrust of other people*.[26]

One of the most theoretically fruitful consequences of this teleological and anti-Kantian system of classification is that it allows us to *compare and contrast the merits of different strategies for diverting these ideas*, just as one might compare and contrast the merits of different automobiles or apartments before committing to one: "The ego gains much less advantage from choosing *transposition* of affect as a method of defence than choosing the hysterical *conversion* of psychical excitation into somatic innervation. The affect from which the ego has suffered remains as it was before, unaltered and undiminished, the only difference being that the incompatible idea is kept down and shut out from recollection."[27] We see here glimpses of Sullivan's later system in which, of all the forms of schizophrenia—the catatonic, the hebephrenic, the paranoid—the hebephrenic is the *most inferior strategy* because it results in a total break from society; the catatonic is preferable because there is at least the hope that, at the culmination of the catatonic retreat, one might learn how to integrate dissociated contents back into the personality. It is as if the mad person has surveyed, in advance, all of the possible forms of madness, as one would contemplate potential romantic partners, and selects the one that is most compatible with her personality.

This anti-Kantian system of classification forms the intellectual framework of the first edition of the *Diagnostic and Statistical Manual of Mental Disorders* in 1952, the beginning of a long and very successful franchise of the American Psychiatric Association (APA). The *DSM-I* divides up all mental disorders into two classes, the organic and the non-organic; it further divides the non-organic into three classes, the psychotic, the neurotic, and the personality. In the fairly brief descriptions that the DSM-I gives of these general categories, *each type is presented as a distinctive strategy for coping with stressors.* This is a fundamentally Freudian insight. The *psychotic* patient copes with stressors by withdrawing from reality:

> A psychotic reaction may be defined as one in which the personality, in its struggle for adjustment to internal and external stresses, utilizes severe affective disturbance, profound autism and withdrawal from reality, and/or formation of delusions or hallucinations.[28]

The *neurotic* patient achieves the same end but through conversion or displacement:

> Grouped as Psychoneurotic disorders are those disturbances in which "anxiety" is a chief characteristic, directly felt and expressed, or automatically controlled by such defenses as depression, conversion, dissociation,

displacement, phobia formation, or repetitive thoughts and acts. . . . [T]he personality, in its struggle for adjustment to internal and external stresses, utilizes the mechanisms listed above to handle the anxiety created.[29]

Finally, the personality disorders are "those cases in which the personality utilizes primarily a pattern of action or behavior in its adjustment struggle, rather than symptoms in the mental, somatic, or emotional spheres."[30]

Moreover, each of the main disorders is, when possible, subdivided into a number of specific "reactions"—for example, the category of the psychoneuroses is subdivided into "anxiety reaction," "conversion reaction," "phobic reaction," and so on. The term "reaction" is Adolf Meyer's expression for a coping mechanism for dealing with an unpleasant situation, a kind of strategy or contrivance for navigating the trials of life. Meyer insisted that the psychiatrist understand mental illness as a goal-directed reaction to the demands of life *rather than an expression of an underlying pathology*:

> The concept of substitutive reactions is meant to keep us from wandering from the ground of the experimental formula of investigation. To try and explain a hysterical fit or a delusion system out of hypothetical cell alterations which we cannot reach or prove is at the present stage of histophysiology a gratuitous performance. To realize that such a reaction is a *faulty response* or *substitution of an insufficient or protective or evasive or mutilated attempt at adjustment* opens ways of inquiry in the direction of modifiable determining factors and all of a sudden we find ourselves in a live field, in harmony with our instincts of action, of prevention, of modification, and of an understanding, doing justice to a desire for directness instead of neurologizing tautology.[31]

And while the *DSM* does not specifically mention Freud, it does tell us that one of its goals is to synthesize earlier systems of classification (in particular, the brief system given in the *Standard Classified Nomenclature of Disease* of 1933) with "the concepts of modern psychiatry and neurology"—that is, of course, psychoanalysis. As we will see, in order to understand this franchise that we call the *DSM* in its multiple editions, we must recognize that the successive editions of the *DSM*, particularly the *DSM-II* and the *DSM-III*, represent attempts to eradicate teleology: their goal is to institute a collective forgetting.

12

Madness as Creativity and Conquest

Kurt Goldstein adopted Freud's teleological orientation and *biologized* it, just as Griesinger biologized Kant. He showed that the forms of mental illness represent the organism's attempt to transform its environment in order to avoid the threat of greater catastrophe (Section I). Goldstein began the study of mental illness with an overarching, philosophical, "holistic," and teleological understanding of what the organism itself is (Section II). In the context of this holistic vantage point, he demolished the very idea of the reflex and warned against our tendency to mistake a teleological performance of the organism for a dysteleological one (Section III). It is true that Goldstein recognized the role of dysfunction in madness—many of the war veterans he treated had traumatic brain injuries—but he taught that, *in addition to* the inner dysfunction, we must, at the same time, see the illness as part of the organism's project of self-actualization (Section IV). We can finally understand the relationship between Goldstein and Freud: despite their differences, both accepted, at a very abstract level of analysis, the principle that madness is always a strategic response to a potentially catastrophic situation (Section V).

I.

Freud taught us how to see, in the ravings of the mad, in the anxieties of the phobic or hysterical patient, in dreams, in slips, what he called a *psychopathological structure*. The idea of a psychopathological structure evokes the image of a well-crafted artifact. On the traditional, Hippocratic conception of madness-as-dysfunction, the very idea of a psychopathological structure is a contradiction in terms: the *pathological* is precisely the absence of design, the breakdown or deformation of structure. But for Freud, the disease is not an aberration, a missing of a target, but a way of achieving a target. And this, in a double sense: the structure is also what he calls a *compromise*. It is this joint endeavor of two different parties, each of which has a life-or-death stake in

Madness. Justin Garson, Oxford University Press. © Oxford University Press 2022.
DOI: 10.1093/oso/9780197613832.003.0013

the outcome. For the *unconscious system*, the structure is a wish fulfillment; for the *ego*, the structure fulfills the wish in a manner that is acceptable to consciousness. One can think of two different agents screaming two different things: one says, "Fulfill this wish," the other says, "Don't traumatize the ego." The psychopathological structure satisfies both demands.

Goldstein also helps us to rethink disease as a whole outside of dysfunction and squarely inside of teleology. A disease is constructed, as it were, from two different ingredients or elements. The first is an actual limitation. In his wounded war veterans, this is a brain injury. The brain injury is debilitating; there is a real dysfunction, a *shrinkage* of ability. But this is not the final word. We must also see in the disease a compensation, a manifestation of what he calls the tendency toward the self-actualization of the organism (*Selbstverwirklichung des Organismus*). In his brain-injured soldiers, whom he ultimately viewed as the paradigm of disease as a whole, we look at each symptom and ask, not, or not *only*: *how does this symptom—for example, that they write in very small letters or have a compulsion toward orderliness—reveal the limitation, the shrinkage, of their world, and the inability wrought by the brain damage?* We also ask: *how is this symptom an expression of the self-actualizing tendency of the organism?* Like Freud, we see the disease within a two-dimensional framework. In Goldstein, it is true, there is a dysfunction, there is a *dis*-ability, but we never use that as the sole principle of disease. Every symptom has this dual quality, a shrinkage and a compensation. In short, *disease represents the organism's tendency to self-actualization, operating within a more limited jurisdiction than usual.*

An analogy might help: you have been given the role of president of an organization, but you are unable to fulfill the duties of your job, and you are demoted to vice president, where you perform perfectly well. When someone describes the kind of work you are doing, one way to describe it would be to say, "This is the kind of work that someone who is unable to be president would do; this is what happens when someone is unable to have a greater jurisdiction, a greater scope of responsibility." One could also say, "This is someone who is producing excellent work within a more limited realm of responsibility or jurisdiction."

Goldstein gives us a typical example that he drew from his Institut zur Erforschung der Folgeerscheinungen von Hirnverletzungen (Institute for Research into the After-Effects of Brain Injury), in Frankfurt, of a brain-injured patient who could not tolerate the least amount of disorderliness in his environment. If a pencil and paper were set in front of him, he would have to

align the paper with the edge of the table, and then orient the pencil parallel
to the paper:

> If, without comment, I again set the pencil obliquely on the paper, the pa-
> tient, provided he has been watching, may once more place it in the same
> way as before. This game can be repeated several times, until he is either
> distracted by something else, or is told explicitly that I want it this and this
> way. In this case, the patient resigns himself to the situation, though usually
> with an expression of marked discomfort.[1]

One rather naïve way to describe his obsessive behavior is to say that it is
an expression of a brain dysfunction. That is not, however, how Goldstein
describes it. Rather, as a result of the brain dysfunction, the patient is extremely sensitive to the experience of having too many possibilities—to use
a more modern term, too many affordances—in his surrounding world;
this experience creates a sense of anxiety; he manages his anxiety by organizing his environment in such a way that each object, as it were, proclaims
its function, proclaims a univocal use. Thus, we must see in the obsessional
behavior not merely a dysfunction but a purpose being fulfilled. Suppose an
artifact is positioned in such a way that it is not clearly disposed to perform
its duty, or that it is disposed to perform an anomalous duty: a can of screws
is being used as a doorstop, or a toothbrush as a pot-scrubber. The brain-injured veteran feels invaded by a multiplicity of possibilities; he must restrict
the affordances of the object to a single one as a way of managing his anxiety.

This example is useful not only in that it reveals the two-fold character
of disease, this psychopathological structure, but also because it shows the
disease to be, at root, a pattern of interaction between organism and environment, the organism's strategy for managing its interactions with the environment. For Freud, the psychopathological structure is, in the first place, an
expression of the mind's attempt to manage its internal affairs well. The mind
must arrange its own contents in such a way as to minimize the chances of
unpleasure, and it does so by carrying out a complicated and fragile negoti-
ation among the inhabitants of its inner world. For Goldstein, too, the activities of the brain-injured veteran must also be understood as a complicated
strategy for minimizing anxiety. In this, Goldstein and Freud are one. But
these activities succeed, in the first place, *not* by managing the contents of the
mind, by forming "compromises" between the mind's faculties, but by struc-
turing the environment in such a way that anxiety-provoking situations are

not permitted to arise. So, while Goldstein says a lot of things against Freud, and makes a lot of "noise" about how he does not agree with Freud, how his view is quite different from Freud's, how he rejects Oedipus, the death drive, the pleasure principle, etc., we must see Goldstein as working within a similar framework, even instantiating a formula that, at a sufficiently high level of abstraction, is identical: the symptoms of the disease represent attempts on the part of the individual to avoid displeasure. It would be better to say not that Goldstein repudiated or opposed Freud, but that he *biologized* Freud. He set this Freudian formula, madness as a strategy for minimizing displeasure, in the context of the organism-environment relationship, within a biological and ecological standpoint. And as we will see, Goldstein goes to great lengths to rediscover the Freudian unconscious *within* the changing relationships between organism and environment.

II.

But what does it mean to "biologize" Freud? What does it mean to biologize anything? As we have seen, different thinkers have different conceptions of what the *bios* amounts to, what the study of life is. This is because they have different understandings of the nature of living matter. As we have seen, Pinel, in invoking the *vis conservatrix*, is not invoking a special vital principle superadded to matter; he is simply articulating what he understands nature itself to be: nature is *essentially* a teleological, goal-directed process. Goldstein, too, repeatedly insists that he is not a vitalist; he violently repudiates any hint of mind-body dualism; he is allergic to dualism. His view of disease stems from a deeper view of what the organism is and what it is up to. Disease must be understood from the standpoint of the organism going about its business in the world. What, then, *is* the organism's business?

Goldstein's most important work, of course, is his 1934 *Der Aufbau des Organismus: Einführung in die Biologie unter besonderer Berücksichtigung der Erfahrungen am kranken Menschen* (The structure of the organism: Introduction to biology with special consideration of the experiences of the sick). He wrote this book in a boarding house in Amsterdam over a six-week period after his exile from Germany. Five years later, as professor of neurology at Columbia, Goldstein oversaw the production of the English translation of the book. The 1939 translation significantly expands on central themes of the book, in particular, his conception of self-actualization and his intellectual relation to Freud. The English title alone indicates a new emphasis on the holistic character of the text: *The Organism: A Holistic Approach to Biology Derived from Pathological Data in Man.* It is not simply a translation, then, but a doctrinal flowering, and for this reason I draw primarily from the 1939 English edition in what follows.

The Organism is celebrated today as a landmark of twentieth-century physiology; in commentaries today, it is remembered for its radical anti-localizationist stance, its repudiation of the concept of the reflex, and its plea for a holistic reorientation of biology and psychology. Such encomiums are adequate if our goal is to rattle off a list of talking points about Goldstein's book, if our goal is to "situate" his work in something like a historical narrative of twentieth-century physiology and psychology. But even before we present a series of talking points about the book, we must ask ourselves, *what is this book about?* And that in a two-fold sense: not only, *what is the topic of the book*, but also in the sense of, *what is the book up to? What is it trying to do? How must one read this book?*

There are two things the book announces right at the outset: its *philosophical* character and its *holistic* character. First, though the 1934 edition calls itself an introduction to biology, this is not a *textbook* about the organism. This is not a compendium about the different aspects or features of the organism. A textbook would have many different chapters, in which each chapter describes one system—respiration, circulation, digestion, reproduction—and all of the chapters collectively yield a panoptic vision of what the organism is. Goldstein would have had no business writing such a book; he dictated its contents in a boarding house in Amsterdam; he did not have a proper lab; he did not have a fixed academic appointment; he was largely destitute. Rather, the purpose of the book is what we often call "philosophical," or even "metaphysical," or "ontological"; it does not obviously fall under the jurisdiction of science, but remains one step outside of it, lurking somewhere around its perimeter. His book means to tell us *what the organism is.* When we recall that Goldstein had originally planned to study philosophy before going into medicine, it is not particularly surprising that he wrote what we should understand to be a philosophical monograph on the nature of the organism. His concern was with basic ontology because he believed that the basic ontology that physiologists relied upon to approach the organism was flawed in such a way as to produce systematic error; once we correct our fundamental ontology, certain facts will come into focus in their true light, their true ordering and significance.

Second, the book has a *holistic* character. He is not writing, in the first place, about this or that feature of the organism. We are not trying to understand one or another *aspect* of the organism; rather, he wants to introduce his conception of what he repeatedly calls "the organism as a whole" (*des ganzen Organismus*). True, certain chapters deal with specialized themes: theoretical reflections on the nervous system; the nature of partitive processes; modification of function due to impairment. But we must not read those chapters in a piecemeal or fragmented way, but as different perspectives on the organism as a whole. It is not a question of breaking an object into smaller parts and interrogating the parts; it is one of slowly turning an object around and looking at it from different angles; nothing must be reified, nothing pulled apart:

> In light of our approach, the problem of the interaction between mind and body appears in entirely different aspect. . . . We are always dealing with the activity of the whole organism, the effects of which we refer at one time to something called mind, at another time to something called body. In noting an activity, we describe the behavior of the whole organism either through the index of the so-called mind or through the index of the body.[2]

But what *is* holism? Holism, for Goldstein, is not just the doctrine that "the whole is more than the sum of its parts," that is, the thesis that studying all of the different parts of the organism separately will never add up to an understanding of the whole. It is not merely that Goldstein was devoted to a "holistic" *rather than* a "mechanistic" understanding of the organism, or an "anti-reductionist" rather than a "reductionist" understanding. And, in fact, historians do say things like this: "Goldstein belonged to the chapter of biology known as German holism." It is true that Goldstein often cites, approvingly, scientists who, according to the historians, have a "holistic" orientation or who "belong to the school of German holism," such as Cannon or von Uexküll, or the Gestalt psychologists such as his friend Karl Lashley. The problem is that *we do not yet know what holism is*. Consider a textbook characterization of holism: "Holism claims that one cannot understand organisms merely by analyzing their various parts, because the parts act differently when they are isolated in the lab than when they are interacting with one another and with the outside world." This neat definition involves a fundamental error: for Goldstein, the problem is not that the whole is more than the sum of its parts, but that, in a sense to be determined, *an organism has no parts*.[3] Parts *exist* only in the context of pathology or disease. Parts *exist* only when a particular animal that we refer to as an "organism" is in crisis. What often enough transpires in a lab under the rubric of physiology, when we isolate parts and try to manipulate them, when we seek to discover "invariances under intervention," is that we are *inducing pathologies*; we are replicating the process of nature whereby an organism gets hurt, and *only then* does it really have a severed limb, or a muscle dangling from the end of a nerve, or a "survival instinct." *A healthy organism does not have a "survival instinct."* A survival instinct is something that *comes into being* when the organism is in crisis.[4] For Goldstein, this is what holism means. It is not a rejection of reductionism but a rejection of mereology.

These two aspects of Goldstein's treatise, its philosophical character and its holistic character, are not independent of each other. For what we discover is that when we look at the organism from this philosophical perspective, when we ask, as it were, about the *nature* or the *essence* of the organism, we come to understand that only a holistic perspective is appropriate or adequate to our subject matter. That is, once we understand the nature of the organism—which we have yet to do—we see the necessity of adopting a holistic vision of it. It is not as if, once we understand what the organism is, we will appreciate that there are two legitimate and even complementary perspectives one can adopt: the *holistic* perspective and the *reductionistic* perspective. The latter necessarily misrepresents the object of knowledge.

What, then, is the essence of the organism? For we need to have knowledge of the organism's essence in order to have a guideline, a *principle of selection*, for determining, say, what facts should be included in a textbook of physiology and what facts should be left out of it. *The Organism* is, in this way, a prolegomenon to a textbook of physiology:

> The analysis of a variety of phenomena has strongly impelled us to the holistic view of the organism. Yet it has not furnished a decisive stand regarding a substantiated knowledge of the structure of the organism. . . . We still cannot give account why we regard precisely these phenomena as essential traits of the organism. We need guiding lines which permit us to make systematic determinations. We need a criterion that enables us to select from the multitude of observations those facts that are suited for the determination of the real nature of an organism.[5]

He then goes on to give us a preliminary (though ultimately, as he realizes, incomplete) statement of this essential characteristic: the essence of the organism is the striving for self-maintenance:

> The criterion as to whether a single phenomenon is such a characteristic of the organism, is, we believe, given in the fact that it is an intrinsic factor in the maintenance of the relative *constancy of the organism*. In contrast to the diversified and even contradictory character of the partitive data, the organism proper presents itself as a structural formation which, in spite of all the fluctuations of its behavioral pattern in the varying situations, and in spite of the unfolding and decline in the course of the individual's life, retains a relative constancy.[6]

This is only a preliminary statement. His statement of the essence of the organism will still have to be modified. The problem with this way of putting it is that it suggests that some principle of homeostasis, of mere survival, even *persistence of form*, is the essence of the organism, but Goldstein does not actually think this. Persistence of form is too paltry a principle for capturing the basic activity of life. Rather, the principle of *self-actualization* supplies us with the essential activity of the organism.[7] Still, the importance of this preliminary formula is that it draws out the holistic character of Goldstein's vision; it draws out the fact that once we understand the nature of the organism, we will appreciate the need for a holistic approach to it; we will avoid the temptation to look at this or that function but rather keep our sight on how the organism perpetuates its existence over time.

In regard to its holistic character, the amoeba actually gives us a much more correct, much *truer*, portrait of the organism than a human being. And this is as it should be: for if we have captured the essence of the organism, we have captured a necessary feature, a feature that we can see in every single instance of what we call "organism," and the thing that makes organisms *organisms*. So, from the point of view of a book called *The Organism*, to the extent that it is a philosophical monograph on the nature of the organism, the study of the amoeba should yield the same results, should convey the same truths, as the study of the giraffe. In a sense, the protist is an organism stripped to its fundamental elements, nature's thought experiment. The reason the protist is a better model for thinking about the organism is that its various activities—movement, digestion, budding—strike us, when observed through a microscope, not as under the jurisdiction of separate faculties with their own functions, but rather as an undifferentiated organism-level reaction to the environment. An untutored spectator cannot see, in the phenomenon of an amoeba digesting a paramecium, the operation of distinct faculties with distinct jurisdictions. Rather, the whole amoeba opens itself out, encircles the paramecium, and *assimilates* it. And although a scientifically trained observer might speak in this faculty-centered way ("locomotion in the amoeba is a function of the pseudopods") the untutored mind cannot see anything but a coordinated movement of the entire organism, a mass action of the whole.

When it comes to human beings, it is much easier, more "natural," to say things like "Eating is mainly a function of the digestive system," "Walking is mainly a function of the neuromuscular system." And there is some truth in that; the human being is much more highly differentiated than a protist. It is because of specific facts about evolution, about how humans evolved, that it is sensible and reasonable to attribute various organism-level activities, such as eating or movement, to specific systems. But we must do so in such a way that the image of the whole organism, the image that seems so intuitive in the case of the amoeba, is always superimposed onto this more differentiated, faculty-centered vision. We cannot lose sight of the fact that every performance is a performance of the whole organism.

III.

This whole-organism standpoint is the basis for Goldstein's attack on the concept of the reflex. For the idea of the reflex is the mythical idea that a certain activity of the organism can be, as it were, completely explicated in terms of something like a distinct biological pathway. Goldstein's is not so much an anti-localizationist doctrine as it is a doctrine that opposes the possibility that there could ever be a scientifically adequate explanation of a single performance of the organism that cites only some specialized part or feature or pathway or system or mechanism. Consider the knee-jerk or patellar reflex, referred to as a "monosynaptic" reflex because it deploys only two neurons with a single synapse between them: the afferent sensory neuron, which conveys information about the quadriceps muscle, and the motor neuron, which causes the muscle to contract. One reason it is such a paradigmatic reflex is that it does not even involve the so-called higher cortical centers. It would seem to be mediated entirely by the lower spinal cord. On further reflection, however, it becomes obvious that this way of describing the reflex is a mistake: there would be no "patellar reflex" until, first of all, the individual feels compelled or persuaded to sit down in a chair, until she agrees to hold the rest of her body relatively stable and tense, until she agrees to allow her lower leg to pivot without resistance, etc. Contrary to initial appearance, philosophical rigor demands that we speak of the patellar reflex as a whole-organism phenomenon and even a "higher cortical" phenomenon, as a phenomenon that "recruits" the higher cortical centers. *The patellar "reflex" involves the entire organism, mind and body.*

This is the moment at which we can expect to hear the tired protest, the irked response, from the physiologists, from the neuroscientists: *of course we know all those things about the reflex. To the extent that we focus on these two neurons, it is not because we think these two neurons could activate the patellar reflex in a vacuum. The only reason we do not mention the engagement of the body as a whole, the higher cortical centers, and so on, in bringing about the reflex, is that it is perfectly evident; it is implicitly understood by everybody that the reflex involves the coordinated movement of the whole body; it is a whole-organism phenomenon.*

Philosophers of biology are well acquainted with this protest; we are familiar with it from parallel debates about genetic causation.[8] Theorists such as Susan Oyama and other proponents of "developmental systems theory" routinely insist that genes do not *cause* phenotypes; the phenotype emerges as a result of interaction between gene and environment. The boring rebuttal to developmental systems theory is that *everyone knows* that a gene for obesity or dyslexia does not give rise to its corresponding phenotype in a vacuum. But as Oyama has forcefully shown, this refrain, *everyone knows that—everyone knows that the genes require an environment in order to be manifested, nobody needs to be told that anymore*—this insistence is usually accompanied by an egregious error: the environment is acknowledged, then quickly relegated to a kind of necessary but undifferentiated backdrop; it has nothing specific, nothing creative to add to the phenotype. It is instructive here to remember Aristotle's view that all of the "information" for the organism, the form-giving principle, is buried in the sperm; the egg is merely the bland background condition for the manifestation of that potential. Likewise, when geneticists say "We *know* that all phenotypes result from the interaction of gene and environment," they often, and at the same time, refuse to attribute "causal specificity" to the environment; the environment is the egg in Aristotle's biology. Put differently, this tired refrain,

everyone knows the environment is needed for manifesting the gene's phe-
notype, nobody needed a philosopher to tell them that, often enough
presupposes that the phenotype somehow *belongs* to the gene; it is *latent* in
the gene; the gene possesses all of the potentiality; the phenotype is *enabled*,
but not *co-constructed*, by the environment. It is in the same manner that we
should read Goldstein's repeated insistence that the monosynaptic pathway
described in neuroscience textbooks does not cause "the reflex"; it does not
have, as its proper activity, the "patellar reflex"; the "patellar reflex" is co-
constituted by the entire body and mind.

Why is this such an important lesson? Why must we repeatedly insist that,
strictly speaking, the traditional idea of the reflex is incoherent? Similarly,
why must we insist, repeatedly, that gene and environment co-construct the
phenotype? The reason is that, *if* the environment is to be more than just a
boring backdrop for the activity of the gene, *if* the gene and environment
are to co-construct the phenotype, *if* the environment carries its own "causal
specificity" relative to the gene, *then* it must be the case that one and the same
gene will produce very different phenotypes in different environments. And
that would decisively undermine talk of there being a "gene for obesity" or a
"gene for depression," if such talk, indeed, is intended to point to the mythical
latency or potentiality of obesity, or depression, in the gene. Of course, if all
such language—e.g., "gene for obesity"—simply means that such-and-such
gene, when interacting with such-and-such environment, "makes a difference" to obesity, or depression, etc., then there is no strict error in such talk,
though it still might be misleading.[9] Similarly, *if* the reflex is co-constituted
by the neural pathway and the rest of the body together, *if* the reflex is not "latent" in the pathway, then one and the same neural pathway will produce different reflexes in different contexts. And that has a bearing on how we lump
together different reflexes, that is, how we classify or *individuate* reflexes.
We may have to subdivide reflexes differently depending on the different
contexts in which they are elicited.

This is, in fact, the basis for Goldstein's critique of the Babinski reflex. The
Babinski reflex is thought of as "normal" in infants. If you stroke the bottom
of the foot, the big toe moves upward and the other toes fan out. After infancy, it disappears in normal adults: if you stroke the bottom of the foot,
the toes move inward instead of upward. The disappearance of the Babinski
reflex is thought to reflect the correct myelination of the corticospinal tract.
If the Babinski reflex, however, resurfaces in an adult, that is a sign of nerve
damage. But this very conventional definition of the reflex raises a profound

philosophical question: what makes that performance, that particular behavior, in an infant, and that behavior, in an adult, instances of "one and the same" reflex, one healthy, one disturbed? By what right do we say that this particular performance *in an adult*—big toe moves upward and others fan out—is an instance of the "Babinski reflex," one that is temporally misplaced? According to our standard vision of physiology, it is one and the same reflex because it has "one and the same" cause; the various instances are due to one and the same distinct biological pathway. The problem that Goldstein identifies is that when we see that performance in an adult, say, in our brain-injured veteran, it is actually an instance of the *flight response*. If a reflex is only a reflex because of the relationship in which it stands to the rest of the body, we have no a priori right to say that both instances of the behavior—the behavior in an infant and in an adult—are expressions of the same reflex. The so-called Babinski reflex, when it occurs in the brain-injured war veteran, does not exist in the infant.

The Babinski reflex demonstrates another point. It does not just illustrate a precise implication of this holistic approach to the organism for medicine and physiology. It also shows us something of much greater significance for the present context, which is *the ease with which we mistake a teleological performance*, a purpose-driven or functional performance of the organism, with a dysfunction, or *merely* with a dysfunction. The Babinski reflex in adults is typically considered to be the sign of a corticospinal dysfunction. While there is some truth to that, it embodies the same error that permeates this dysfunction-centric perspective on disease. According to Goldstein, when we see the so-called Babinski reflex in an adult, it is not *merely* a dysfunction: it is also an expression of the goal-driven activity of flight from danger. But this raises the question of what the organism's fundamental *goal* is.

IV.

We have stated that, in order to "do biology," in order to know which activities are characteristic of the organism and which are incidental, we have to grasp the organism's *essential activity*. So, once more, what is the essential activity of the organism?

Earlier, we said that the essential characteristic of the organism is maintaining constancy. And though that was valuable for helping us understand the holistic character of Goldstein's investigation, it was inadequate, because it makes it sound as if the organism's main goal is mere survival, persistence, homeostasis. But in fact, Goldstein urges us to see the essential activity of the organism not as mere survival, mere homeostasis, but rather as what he calls the *tendency of self-actualization*. (Maslow, a student of Goldstein, explicitly credited Goldstein for the phrase, which he made famous in his "hierarchy of needs," though in a perverse, atomized form.)[10] The reason we must get a clear picture of self-actualization as the essence of the organism is that it will give us the principle of disease. *Disease is not the stunting or arresting of the drive for self-actualization but the exercise of that drive in a more limited sphere or jurisdiction.* One implication of this is that disease is not a deviation of the organism from its essential nature. It is a furtherance, a continuation, an expression, of that very nature.

Goldstein's primary discussion of self-actualization is situated in the context of a critique of Freud's notion of culture as a sublimation of repressed drives:

> This [Freud's theory] would mean a complete misapprehension of the creative trend of human nature, and at the same time would leave completely unintelligible why the world was formed in these specific patterns. . . . This becomes intelligible only if one regards them as expressions of the creative power of man, and of the tendency to effectuate a realization of his nature. Only when the world is adequate to man's nature do we find what we call security.[11]

So, self-actualization has two interrelated features. The first is that it is the essence of *creativity*, not *survival*. The creative person is the one whose creations serve to articulate her inner nature. But the notion of self-realization has not only the resonance of creativity, but the resonance of conquest, and specifically the *joy* of conquest.[12] Conquest qualifies the nature of the creative impulse he describes. It is a kind of creativity that consists in the transformation of one's environment in such a way as to render it more *in one's likeness*, and to derive pleasure from that resemblance. It is a transformation of the environment in such a manner that the environment is now a mirror of my inner nature. This transforming capacity is not fueled by anxiety; it is fueled by joy.

Now, with this picture, it would be tempting, though mistaken, to say that, with the brain-injured person, there is a stunting or even disappearance of the tendency to self-actualization. This is precisely the way Maslow describes the situation in his revolting "hierarchy of needs": when the more basic or "lower" physiological needs are threatened, the "higher" drive toward self-actualization disappears: "If all the needs are unsatisfied, and the organism is then dominated by the physiological needs, all other needs may become simply non-existent or be pushed into the background."[13] This makes it sound as if, in disease, the tendency to self-actualization is strangled because the environment is overwhelming and the individual is not "up to" its demands. But this is a mistake: Goldstein, in contrast to Maslow, specifies that "even the brain injured patient is not permanently in a state of anxiety, and we have seen how a transformation and shrinkage of his world spares him from such a condition."[14] The brain-injured patient *shrinks* his world in such a way that he can continue with his self-actualizing project, his creativity and conquest, his attempt to transform the world into a mirror of his nature. Hence, in disease we do not encounter a *lack* or an *absence*, but a *funneling* of the self-actualizing project through a narrower channel.

If we misunderstand this principle, it would be all too easy for us to reinterpret or recontextualize Goldstein's work through a dysfunction-centered lens. To be sure, Goldstein does state that the brain-injured veteran has lost a fundamental psychological capacity, namely, *the capacity for abstraction*. It has been abolished. This is a theme he develops at length in his *Human Nature in the Light of Psychopathology*.[15] The abolition of the capacity for abstraction is a dysfunction. The brain-injured veteran only thinks in terms of the concrete, the specific, the particular. For example, if you ask her to sort a pile of color swatches, say, to remove all of the red ones from the pile, she can do so, but only if she constructs a series of swatches of gradually descending hues. She succeeds in the assigned task but by using a deviant route: she must follow a rigid sequence of descending hues. Put differently, the brain-injured veteran does not "have the concept" of *red*. Still, while she does have a dysfunction, she also has acquired a new ability; she exhibits a kind of ingenious resolution of her problem, a creative conquest of her world. We should speak of her performance, then, not just as a dysfunction but as an instance of human problem-solving working under a novel, and more demanding, set of constraints.

This is precisely the principle that Canguilhem, in his 1966 treatise, identifies, and extols, in Goldstein's work: Goldstein, he says, recognized rightly that "any one act of a normal subject must not be related to an analogous act of a sick person without understanding the sense and value of the pathological act for the possibilities of existence of the modified organism."[16]

Disease is not, or not merely, the deprivation of health and normalcy, *but a new "norm"*: "Diseases are new ways of life."[17]

This idea of disability as a fundamentally altered mode of functioning is echoed in Ron Amundson's manifesto, "Against Normal Function." There, Amundson criticizes philosophical theories of biological function, particularly Boorse's, for invoking the specter of a "universal species design," and for seeing disabilities as so many failures of, or deviations from, this design, *rather than as modes of functioning in their own right*. He talks of Slipjer's goat, a goat that was born without forelegs but which learned to walk upright:

> Slipjer's goat had many other deformities (relative to the statistical norm) in its skeletal and muscular anatomy. It had an S-shaped spine, an atypically broad neck, many atypically shaped bones and atypically positioned muscles. Its thorax was oval shaped, unlike the V-shaped cross section of the typical goat. By this census of 'abnormalities' it was a radical departure from its species design, and each abnormality pulls it further from the norm. By the species design criterion of goal-directedness, Slipjer's goat was a notable failure. By the developmental criterion it was a roaring success. The goat's skeletal and muscular abnormalities were, each of them, adaptively suited for life as a biped. They mimic the body conformation of kangaroos and humans.[18]

We must learn to see, in disease, this fusion of creativity and conquest: disease as *roaring success*.

V.

Disease is not a disruption of the organism's self-actualizing tendency, but a further expression of it. And this takes us directly to the problem of anxiety, and ultimately, to Goldstein's confrontation with psychoanalysis. First, the motor of anxiety: when the organism confronts a situation that it is not "up to," it experiences anxiety. Anxiety is a *subjective* manifestation of a looming, *objective* catastrophe.[19] Fear, for Goldstein, is actually the fear that I am *going* to be anxious; this fear causes me to modify the environment in such a way as to minimize the prospect of this anxiety-provoking catastrophe. Now we can see a fundamental similarity, even unity, between his view and Freud's. Both Freud and Goldstein agree that the mind is constituted, in part, by a series of hedonic devices, devices that strive to minimize displeasure. The difference is that, for Freud, these devices are involved in the mind's project of managing its internal economy; for Goldstein, they are involved in the organism's project of managing its environment. So, despite the critical tenor of Goldstein's remarks on Freud—Goldstein doesn't believe in separate drives; he doesn't believe in a reified unconscious, the pleasure principle, and so on—they share a fundamental formula. If we wanted to be clever, we could say that both Freud and Goldstein agree that these psychopathological structures result from the *individual's attempt to minimize the prospect of catastrophe*, but that they understand "catastrophe" differently: for Freud, minimizing "catastrophe" means preventing the conscious occurrence of an unpleasant idea, while for Goldstein, it means preventing the occurrence of an environmental challenge that the person is not "up to."

Intriguingly enough, Freud seems to recognize precisely the possibility that the psychopathological sequelae of brain injury could be explained by such a generalization of his viewpoint. In 1921, he wrote the introduction to an anthology on war neuroses; of such neuroses he comes ultimately to the following formula:

> In the traumatic and war neuroses the ego of the individual protects itself from a danger that either threatens it from without, or is embodied in the form of the ego itself, in the transference neuroses of peace time the ego regards its own sexual hunger (libido) as a foe, the demands of which appear threatening to it. In both cases the ego fears an injury: in the one case through the sexual hunger (libido) and in the other from outside forces. . . . [O]ne can [therefore] with full right designate the repression which underlies every neurosis, as a reaction to a trauma, as an elementary traumatic neurosis.[20]

What is important to appreciate is that this formula is not an *extension* of Freud's theory but a *generalization* of his theory. What, exactly, is the difference between an extension and a generalization? Freud's theory of dreams, his theory of jokes, his theory of slips, are *extensions* of his theory of hysteria; one and the same *mechanism* generates all of these diverse phenomena; we are simply incorporating new phenomena or new explananda under one and the same explanans. In the case of war neuroses, however, Freud is not *extending* his theory; rather, he discerns that a theory of hysteria and a theory of war neuroses share a certain logical similarity; they are analogues; they share *not a mechanism but a logical form.*

Finally, we are now in a position to state, simply, what it means to "biologize" something, to take a "biological perspective": it is to place the phenomenon in the context of the organism's ever-changing interactions with the environment, *rather than* in the context of the relationship between diverse mental faculties. But the very notion of "biologizing" something undergoes a deep transformation throughout the twentieth century. By the time that we get to the 1980s, to the evolutionary psychologists of today, what we see is precisely the sort of atomization that Goldstein detested; we now understand madness *not* in terms of the goal-directedness of the organism or its tendency toward self-actualization, but the *adaptedness of its parts.* Evolutionary psychology takes us away from the whole organism and shifts our attention to *mechanisms with evolved functions*; it fragments the mind in a way that Goldstein would have found abhorrent. This comes with a corresponding shift in the question of mental disorder: should we see in a mental disorder, such as depression, a dysfunction of a mechanism, that is, the failure of function of an evolved mechanism, or rather, the fulfillment of a function?

13

From Retreat to Resistance

With Gregory Bateson's double-bind theory, and Laing's creative appropriation of it, we glimpse yet a new way of coming to terms with the teleology of madness, for in the double-bind theory, madness is not merely a form of retreat from an untenable social world; rather, it is a way of engaging with it *from a different standpoint*. It is an expression of a steadfast and assertive *nonparticipation* in a corrupt social order (Section I). It should not be entirely surprising, then, that the American counterculture was able to glimpse, in madness, a mirror of its own plight: the mad person had become a social revolutionary (Section II). This brings us squarely to the problem of *acid*: by the Summer of Love, acid became a symbol of this juncture between madness and social revolution (Section III). Extraordinarily enough, these new lines between acid, madness, and social revolution began to transform medical psychiatry in the United States: acid could no longer be seen as a "psychotomimetic drug," but as a tool for heightening one's moral and spiritual awareness (Section IV).

I.

From Frieda Fromm-Reichmann, we derive the image of the mad person as one who has retreated into a private world. She has isolated herself. Isolation is the theme of madness; it is its predominant telos or goal. Corresponding to that, catatonia is the fundamental form of madness; hebephrenia and paranoia are its degenerate forms. Still, as we have seen, this project of withdrawing from the social world is impossible to complete. That is for two reasons: one is more theoretical and the other anthropological. First, in order to enjoy rest in one's self-imposed isolation, one must prevent it from being disturbed or disrupted; if the queen is to sleep soundly, she must ensure that there is a watchman on the wall. This requires a split of the psyche into two functions: the one who is enjoying the isolation, who is resting from the anxiety and pain of rejection, and the one who is vigilant to the social world, and

Madness. Justin Garson, Oxford University Press. © Oxford University Press 2022.
DOI: 10.1093/oso/9780197613832.003.0014

who can adeptly turn back any wanderers, anyone who comes too close to the fortress. Hence Fromm-Reichmann's constant warnings about the power of countertransference and the psychological risk to the therapist. But this—this need for vigilance, the need to turn away potential intruders—is *already a mode of engagement*. Anthropologically, we are human beings who can experience loneliness, and it is easy to think that isolation would be a panacea, that retreat into a castle or refuge, a cabin in the woods, would be better than the messy social world, but because we are human and have a social instinct, we will always experience an unhappy tug toward social living. We will always keep the dream of a perfect social world, one without pain or rejection, alive.

With the double-bind theory associated with Bateson and his colleagues but inspired by Fromm-Reichmann—Bateson and Fromm-Reichmann collaborated during her fellowship at Stanford's Center for Advanced Study in the Behavioral Sciences during the 1955–56 academic year—this facet of madness, its character *as* mode of engagement, is brought to the fore. Not only does this vision reach its zenith, but in the hands of R. D. Laing it is transformed. The mad person is de-Oedipalized and radicalized; he is an agent of social change; he is, in fact, a revolutionary. Madness no longer consists in an infantile flight from the pain of rejection; it is a *refusal to participate* in a corrupt social order. And, as a necessary adjunct to this transition, we no longer think of madness as merely a flight but as a style of revolt. But this identification of the madman and the revolutionary, which was developed by the Socialist Patients' Collective at Heidelberg and, to a lesser extent, Psichiatria Democratica in Bologna, elaborated in Deleuze and Guattari's *Anti-Oedipus*, and perhaps carried on today by movements such as the Icarus Project, is an outgrowth of, is rooted in the soil that was tilled by, Heinroth.

The double bind theory. A boy or girl is repeatedly placed in a special kind of position, the *double bind*, in which any coherent response will be penalized. A mother is unable to show affection, and so the boy does not show affection toward her. She says, playfully, "Don't you love me anymore?" If he responds with affection, she will retreat, and he will experience the pain of rejection. But if he does not respond with affection, her accusation is confirmed, and he will be reprimanded. Therefore, her seemingly innocuous, even playful, question, "Don't you love me?," has placed him in a double bind: anything he does will be punished. All of the elements are in place for the schizophrenic withdrawal, which we are familiar with. There are actually three forms that this withdrawal can take, corresponding to the three forms of schizophrenia: the paranoid, to always look for hidden messages behind normal communication; the hebephrenic, to laugh at every form of communication; and the catatonic, to ignore them. "If an individual does not know what sort of message a message is, he may defend himself in ways which have been described as paranoid, hebephrenic, or catatonic."[1]

One might be tempted to see, in this, merely an extension of Fromm-Reichmann's ideas; to be sure, Bateson and his colleagues were explicit in their indebtedness to her work. Still, there is a fundamentally new theme here, or perhaps an amplification of a theme that was only latent in Fromm-Reichmann: in addition to seeing in schizophrenia a withdrawal from the world, we also see in it *a mode of engagement*. Schizophrenia is a way of engaging with mother, *but from a different standpoint*, a standpoint that does not lend itself to victimization. The individual is not necessarily retreating from the battle scene but changing positions, changing his vantage point: it is a tactical maneuver.

Bateson and his colleagues still stay within the relatively safe haven of the family. This is still a fundamentally Oedipal picture. My "issues" are still mommy and daddy issues. Laing transforms that picture in a decisive way: he de-Oedipalizes it. *Yes*, he tells us, *schizophrenia begins with the family, but the family is merely an instrument of socialization, and it is really the socialization process, that sick socialization process that churns out mindless soldiers, that churns out the Vietnam War, that churns out "better dead than Red," that the mad refuse to participate in*. And we can watch Laing's work of de-Oedipalization unfold in the following passages. He begins by identifying the source of one's problems as problems with mommy and daddy:

> It is only in the last ten years that the immediate interpersonal environment of 'schizophrenics' has come to be studied in its interstices. . . . At first the focus was mainly on the mothers (who are always the first to get the blame for everything), and a 'schizophrenogenic' mother was postulated, who was supposed to generate disturbance in her child. Next, attention was paid to the husbands of these undoubtedly unhappy women . . . then to the nuclear family group of parents and children.[2]

But then he notes that, since mommy and daddy are mere puppets of the socialization process, the scope of the analysis must balloon out to encompass the larger social field:

Questions and answers have so far been focused on the family as a social sub-system. Socially, this work must now move to further understanding, not only of the internal disturbed and disturbing patterns of communication within families, of the double-binding procedures, the pseudo-mutuality, of what I have called the mystifications and the untenable positions, but also to the meaning of all this within the larger context of the civic order of society—that is, of the *political* order, of the ways persons exercise control and power over one another.[3]

And finally, the mad person is someone *who has said no*:

To regard the gambits of Smith and Jones [two people diagnosed with schizophrenia] as due *primarily* to some psychological deficit is rather like supposing that a man doing a handstand on a bicycle on a tightrope 100 feet up with no safety net is suffering from an inability to stand on his own two feet. We may well ask why these people have to be, often brilliantly, so devious, so elusive, so adept at making themselves unremittingly incomprehensible.[4]

With this move, with this gesture, Laing opens up the core intellectual inversion of the anti-psychiatry tradition, and its famous two-part formula:

Sanity is madness

and

Madness is sanity

We can study each part of this formula separately. First, *sanity is madness*: the "sanity" of society, the sanity of our educational institutions, our churches, our fraternities, our boys' clubs, the sanity that leads to the concentration camps and to Vietnam, *that is pure madness*: "Long before a thermonuclear war can come about, we have had to lay waste to our own sanity."[5] The second part of our formula is that *madness is sanity*: the schizophrenic, the one who is fleeing, is fleeing society because she glimpses, even inchoately, its madness, and she rejects its madness: "[Future men] will see that what we call 'schizophrenia' was one of the forms in which, often through quite ordinary people, the light began to break through the cracks in our all-too-closed minds."[6]

Laing uses a simple metaphor to characterize the anti-psychiatry inversion: imagine a formation of jets that are all moving in the wrong direction; they are traveling east rather than north. One of the pilots recognizes the error and changes course, traveling north. The lone pilot is out of formation, but not off course.[7] The analogy to madness is obvious: to be "out of formation" is to be deemed mad; to be "on course" is to be moving in the right direction, the good direction, the morally upstanding direction, the direction of authenticity, love, truth, peace. And it is because of this that Deleuze and Guattari say of Laing that, with the exception of Jaspers, he came closer than anyone else to the de-Oedipalization and radicalization of schizophrenia; he came closer than anyone else to grasping the revolutionary potential of madness.[8]

II.

In retrospect, it was only a matter of time before the American counterculture could see, in schizophrenia so construed, a representation, a kind of *miniature*, of its own plight. And that is because during the 1960s counterculture, we have the idea that one legitimate strategy of political resistance against fascism, against violence, against war, is to adopt an *assertive, expressive, and steadfast refusal to participate*. Fascism thrives on consensus and participation. Fascism cannot impose itself on the masses without their implicit agreement, without their "buy-in." This philosophy of non-participation was poignantly expressed by John Lennon and Yoko Ono's Bed-Ins for Peace. But the mad person, too, Laing tells us, simply embodies this prerogative of non-participation; she enacts this assertive refusal to participate in a more intense and focused way. In fact, there are long passages in *The Politics of Experience* in which we simply *do not know* whether Laing purports to describe the mad person or the global counterculture.

The mad person is the hierophant of this new order, of the resistance to fascism, and the forerunner of a new way of doing things. But this raises a problem: the mad person, to put it gently, does not seem like much of a revolutionary. A disheveled man huddling in a corner of the back wards of St. Elizabeth's Hospital, a woman covered in her own feces on a hospital bed in

Byberry, do not exactly strike us as the torchbearers of the revolution. So, we have a contradiction: the mad person is the revolutionary and he is also the one huddling in the back ward of the asylum, drooling on his beard, or the one covering her body with her own shit. How do we reconcile these two pictures? *We must postulate that there are, in fact, two forms of madness*: there is the true schizophrenia and the false schizophrenia. The true schizophrenia is pure revolutionary zeal, whose modus operandi is a refusal to participate in fascism. The false madness is the disheveled man drooling on himself in the back wards of St. Elizabeth's. It is thus that Deleuze and Guattari, in 1972, identify a split within madness, a true and a false schizophrenia. The true schizophrenic is an utterly exalted figure: he has exploded all polarities inside of himself; he is Schreber; he is the sun, the moon, a man, a woman, an animal. He has refused the most intimate incursion of fascism into our psychological lives; he has refused to adopt a binary existence, *tinker, tailor, soldier, spy*:

> [The schizophrenic] is and remains a disjunction: he does not abolish disjunction by identifying the contradictory elements by means of elaboration; instead, he affirms it through a continuous overflight spanning an indivisible distance. He is not simply bisexual, or between the two, or intersexual. He is transsexual. He is trans-alivedead, trans-parentchild. He does not reduce two contraries to an identity of the same; he affirms their distance as that which relates the two as different. He does not confine himself inside contradictions; on the contrary, he opens out and, like a spore case inflated with spores, releases them as so many singularities that he had improperly shut off.[9]

And in that refusal, he has discovered a new form of life, a new energy, intensity, vitality; his psychic life is dissolved into a play of forces, of desire, of movement, of transgressions, and it is that energy that will spill out into society and that will transform the world. The false schizophrenic, in contrast, is the one who is huddled in the corner, drooling on his own beard, covered with her own feces, on the back wards:

> How is it possible that the schizo was conceived of as the autistic rag— separated from the real and cut off from life—that he is so often thought to be? Worse still: *how can psychiatric practice have made him this sort of rag, how can it have reduced him to this state of a body without organs that*

has become a dead thing—this schizo who sought to remain at that unbear-able point where the mind touches matter and lives its everyday intensity, consumes it?[10]

But this new split, this schism between the true schizophrenia and the false schizophrenia, just raises a new question: what is the relationship between the true and the false schizophrenia? What is the *mechanism* that takes, as input, the true schizophrenia and produces, as output, the false? And the so-lution to the problem was obvious: *capitalism*. Capitalism is the force that takes the true schizophrenic and converts him into the false schizophrenic. Capitalism sees in the true schizophrenic the harbinger of its own demise; it frantically intervenes through a specialized instrument called "medical psychiatry":

> Our society produces schizos [that is, the false schizophrenics] in the same way it produces Prell shampoo or Ford cars, the only difference being that the schizos are not salable. How then does one explain the fact that cap-italist production is constantly arresting the schizophrenic process and transforming the subject of the process into a confined clinical entity, as though it saw in this process the image of its own death coming from within? Why does it make the schizophrenic into a sick person—not only nominally but really?[11]

Capitalism gives her drugs, it gives her electroconvulsive therapy, it gives her a frontal leucotomy, it transforms her into the gibbering lunatic of the back wards, for there she can be maintained; there she is useless. So an inevitable implication, a necessary consequence, of seeing the mad *as* revolutionary is the distinction between true and false madness and the associated idea that fascism is the bridge between them, the function that moves one from the true to the false madness, the revolutionary to the gibbering lunatic, and that medical psychiatry is the agent that implements this function; it is fascism's lap dog.

The thought police are real. They are not a figure, a metaphor, a manner of speaking. There are certain thoughts that a person can have, *and certain be-haviors that would be reasonable enough to perform given those thoughts*, that will trigger a certain social function. Those thoughts and behaviors *are not illegal*, per se, but society somehow recognizes that those thoughts cannot exist without exerting a destabilizing force upon it. It sees that certain kinds of thoughts and behaviors that seem innocent enough, even humorous, simply cannot be allowed to continue in an unsupervised manner without jeopardizing society as a whole. Suppose I wake up, defecate, and I look in the toilet and think, *It would be funny to take a little of that and smear it on my face and go into work that way, everybody would freak out, it's such a little thing and I could even wash it off after, and it would wreak havoc in my workplace, but then I would wash it off and explain myself and everyone would calm down and maybe laugh a little. What would happen if I did that?* Or if I wake up and think, *I want to be a woman, too, I've been told that I'm a man, I have a penis, but I feel like a woman. I ought to put on a beautiful dress and makeup and go into public that way, what would happen?* Or suppose you're an adolescent, and you wake up on a Monday morning and think, *What if I refuse to go to school today? And tomorrow? And the next day? Everyone would freak out, but what would happen, really?*

These are not criminal offenses. Yet they are—or various of them have been seen at various times and places as—socially intolerable. So, we have a system of police, call it the thought police, call it whatever you want, that intervenes and says, *Not only can you not do those kinds of things, but there is something about the way you are thinking that is intolerable. A sane person might entertain thoughts like that, but in a purely hypothetical way; she would never actually act on them; but you have not only entertained those thoughts, but allowed those thoughts to spill over into action.* And that would initiate a series of interventions that could culminate with an involuntary hospitalization, and quite possibly, a whole series of more intensive interventions. So, it is impossible not to see psychiatry as having a social function: in a sense, it is the *cement of society*, in the way that Mackie calls causation the cement of the universe. This is the basic vision of the anti-psychiatrist Cooper. He famously saw in the construction of the mad a pivotal social function being performed:

> Schizophrenia is a micro-social crisis situation in which the acts and experience of a certain person are invalidated by others for certain intelligible cultural and micro-cultural (usually familial) reasons, to the point where he is elected and identified as being "mentally ill" in a certain way, and is then confirmed (by a specifiable but highly arbitrary labeling process) in the identity "schizophrenic patient" by medical or quasi medical agents.[12]

The last clause of this passage is *not* that psychiatry *heals* you, but that it *confirms you in the identity of* a mad person. The real elegance of this system of thought control is not that it makes you stop doing those things. It gives you permission to keep doing them. That is, in a sense, the whole point of it. We are not saying that you have to stop. We are giving you a new option: in addition to the option of stopping, you also have the option of continuing to do it but under a new label, the label of being a schizophrenic person, or a person with bipolar disorder with accompanying psychotic episodes, or a person with a personality disorder, or what have you. That might come with some sort of agreement or certification brokered between us and you, or us and your legal guardians or representatives, for some kind of ongoing supervision.

This new division, this differentiation, between the true and false schizophrenic, implies a new role for the therapist and the healer. Laing sees all of this very clearly. What, then, is Laing's position as a healer? What, precisely, is he being called to do as a therapist and guide to the mad? His role is to *teach the mad how to be mad*. His job is not to cure the sick but to enlighten the mad person about her revolutionary potential and the inexorable societal push to squash this potential. He says that *you* are the superior person, even an enlightened one; *you* are being oppressed because you are leading your people into a new land, into a new political and social organization; *you* need to grasp that role, or you might become a gibbering lunatic in the back ward of an asylum, a neutered non-entity.

Laing takes this basic vision of the good madness—the mad person as the revolutionary who refuses to participate—and he synthesizes it with a spiritual vision of the inner journey. So, there are two wings to his healing process: the mad person, in breaking off from society, is also embarking on an inner journey, and that journey is segmented; there are certain kinds of archetypical challenges that one must confront on that journey:

> This journey is experienced as a going further 'in,' as going back through one's personal life, in and back and through and beyond into the experience of all mankind, of the primal man, of Adam and perhaps even further into the being of animals, vegetables, and minerals. In this journey there are many occasions to lose one's way, for confusion, partial failure, even final shipwreck: many terrors, spirits, demons to be encountered, that may or may not be overcome.[13]

And there are many ways that this journey can go wrong, that the mad person can fail in her quest for transformation, and so what she needs is a guide, and

Laing is one such guide. I have been into the recesses of the inner world; I am an argonaut of inner space. And once you have successfully navigated this inner path, *you*, in turn, will be a leader; *you*, in turn, will shepherd others through that process; *you* will truly be a guide. And indeed, in his Kingsley Hall, the residents were leading the other residents through the journey of healing:[14]

> Instead of the *degradation* ceremonial of psychiatric examination, diagnosis, and prognostication, we need, for those who are ready for it (in psychiatric terminology often those who are about to go into a schizophrenic breakdown), an *initiation* ceremonial, through which the person will be guided with full social encouragement and sanction into inner space and time, by people who have been there and back again. Psychiatrically, this would appear as ex-patients helping future patients to go mad.[15]

But this raises a new question. Why must the mad person be shepherded along this inner journey? Why can I not just explain to the mad person: "You are a political revolutionary and society is trying to oppress and destroy you and now that you know that, you can take on that leadership role"? Why the need for this intermediate step? Why must the mad person embark on an inner journey before assuming this revolutionary role? Is it because of its archetypical resonances: Jesus in the wilderness; Moses on the mountain; the Buddha under a tree? Why the need?

The answer that the counterculture gives is *deprogramming*.[16] The person who only now begins to show prodromal symptoms, the person who is brought by the police to the psychiatric wing of a public hospital, has revolutionary potential, but this revolutionary potential is not yet complete, not yet perfected. Some kind of inner transformation must yet take place to prepare you for that role. And the reason is that society's ideals, its values, still have a strong grip on you; you need a period of deprogramming; you need to withdraw in order to shake off the grip of those categories, ideologies, and habits, and then you can assume this role of the revolutionary, *but not before*, because *before* the inner journey, you have too many social complexes, too many hang-ups, too much ego, to be of any real service.

III.

One cannot think clearly about the relationship between the American counterculture and madness without thinking about acid. Acid is the solid chain, the sturdy link, between the counterculture and madness. By the late 1960s, acid has two profoundly contradictory significations: acid makes you a revolutionary; acid makes you mad. Hence anybody who wishes to rethink the complex interplay of resistance and madness must take a stand on the problem of acid. In the writings of Timothy Leary, or John Lilly, or Allen Ginsberg, acid is not an end in itself. It is a means to an end; its chief function is *deprogramming*. It is said that the reason Leary was imprisoned in over 36 prisons on multiple continents was not because he was a "drug pusher"; it is because everybody tacitly understood that he was an agent of deprogramming and he was freely dispensing instruction on how to do it.[17] For Lilly, we should call acid a "reprogramming substance," rather than a "psychopharmacologically active drug" or "psychotomimetic," terms that carry quasi-medical or moral connotations.

Some today, over 50 years after the Summer of Love, are now celebrating what they see as the "renaissance" of psychoactive drugs like LSD, psilocybin, DMT, in controlled, supervised portions and settings. Enthusiasts of this new tradition praise the potential therapeutic implications of these drugs for everything from depression, anxiety, and weight loss to revitalizing marriages. But it is hard to overstate, from the standpoint of the 1960s, what an abomination this new vision represents. What we have now, what we see today in this "psychedelic renaissance," is a redirection of psychedelic drugs away from any sort of radical or revolutionary potential; their new role is to help people who have been burdened by society to get their lives *back on track*. The so-called renaissance represents the decisive victory of fascism over the individual: it is the 1960s imbibed and regurgitated by the "me generation." It is a way for fascism to declare: *Hey, man, psychedelics are a pretty groovy way of dealing with your awful life, of dealing with the drudgery of the nine-to-five, of losing those extra pounds you can't seem to shake off, of breaking off the bad habits, of putting some pep in your step, of giving you a sense of the cosmic significance of your ridiculous existence.* The slogan and epitome of the psychedelic renaissance is *microdosing*. Microdosing, we are told, is taking trace amounts of psychedelics without disrupting your day-to-day life. Microdosing: get the therapeutic benefits of acid minus the revolutionary zeal! The fact that today it is even possible for us to entertain the idea of microdosing as *something desirable* is a clue to how much our worldview has shrunk, has shriveled, since the aborted revolution.

Acid does not make you mad; it makes you sane. It makes you saner than sane. But immediately, we see a contradiction between the signification given to acid by the counterculture and the signification given it by psychiatry. For the counterculture, acid takes a mad person and makes him sane; for psychiatry, acid does the reverse. Throughout the 1950s and even the early 1960s, acid was widely understood within psychiatry as a psychotomimetic drug, as inducing a "model schizophrenia." So, in the wake of the counterculture, these associations within medical psychiatry have to be rethought. But to understand that, we have to understand, within psychiatry, the early twentieth-century quest for a psychotomimetic drug.

A drug like acid was considered psychotomimetic in two quite different senses, a phenomenological sense and a biochemical sense.[18] In the one sense, it was thought to induce, in a sane person, an experience that allows one to see, "from within," what it is like to be psychotic. It makes you psychotic for a short period of time. This alone gives the psychotomimetic invaluable therapeutic utility because the therapist can now understand the patient from within rather than speculating about the patient's mind in a detached, theoretical way. Psychiatry has experimented with many drugs in the hopes of finding such a psychotomimetic. Kraepelin himself used hash for this purpose but said it only made him tired. Erich Guttmann, William Mayer-Gross, and Humphrey Osmond also used mescaline and hash in addition to acid for those purposes. By the 1950s, there was no question that acid was the ultimate psychotomimetic. Consider a *Time* magazine article from 1955, which tells us that, with acid, "at last [psychiatrists] can turn on and off, at will, psychotic episodes which have most of the earmarks of natural mental illness."[19]

But there is a second sense of "psychotomimetic," in which a psychotomimetic is not merely a drug that induces a psychotic episode in an otherwise normal person, but is a tool for identifying the underlying *biochemical mechanisms* of madness. The reasoning is *similar effect, similar cause*. If acid simulates madness in a normal person, then it is likely to do so by activating the same biochemical pathways that underlie genuine madness itself. Hence LSD had acquired a two-fold role in psychiatry: it was both a therapeutic measure to help the therapist understand the patient's world, and a tool for biochemical research. This second function of acid, as tool for biochemical research, is what gave rise in the early 1950s to the *serotonin hypothesis* of schizophrenia. The serotonin hypothesis claims that schizophrenia is probably due to a deficiency of the neurotransmitter serotonin in the brain, and its justification is that LSD makes you mad; it does so by triggering the biochemical mechanism of madness; when we inject LSD into rodents, it depletes serotonin from their brains; so, madness, too, is probably the result of a serotonin deficiency. Two papers published in 1954 arrived at this conclusion independently. As Woolley and Shaw summarize:

> The demonstrated ability of [LSD and similar ergot-based agents] to antagonize the action of serotonin in smooth muscle and the finding of serotonin in the brain suggest that the mental changes caused by the drugs are the result of a serotonin-deficiency which they induce in the brain. If this be true, then the naturally occurring mental disorders—for example, schizophrenia—which are mimicked by these drugs, may be pictured as being the result of a cerebral serotonin deficiency arising from a metabolic failure rather than from drug action. Possibly, therefore, these natural mental disorders could be treated with serotonin.[20]

One mission of the counterculture, then, and certainly one *effect*, was to subvert this picture of the use of acid as a psychotomimetic drug. *Acid doesn't make you mad, it makes you sane*—or, if we wish to persist in this language of acid "making you mad," we can say that *it makes you mad in the sense of the true madness, where madness is equivalent to a higher sanity.*

But if acid isn't a psychotomimetic drug, if acid doesn't make you mad, what does? The counterculture had an answer to that: speed. If acid is the tool of deprogramming, if acid leads you to take a step back and reflect on the madness of society, if acid prompts the inner journey that leads to enlightenment and to revolution, then speed is its antithesis. Speed is evil incarnate. Speed is America in tablet or liquid form. Speed is fascism's answer to acid. Speed is how fascism says, *if you think acid's pretty far out, you ought to try speed, it'll really blow your mind.* So, what is speed? And not in the sense of the molecular structure of speed, $C_{10}H_{15}N$. I mean, *what is it* in its true nature, its true being?

First, speed was a wonder drug of the 1930s and 1940s.[21] It was doled out en masse to soldiers in England, France, Germany, and Japan, to make them better soldiers. It gave them enhanced focus, enhanced energy, enhanced intensity. After the war, it was widely hailed as a *pick-me-up*, as a mild anti-depressant. It creates a mild sense of euphoria that can help you make it through another grueling day at the office or in the kitchen. So whatever speed is, in terms of its chemical structure, or in its mode of activity in the brain, in its *essence*, speed is the antithesis of acid: speed makes you a *better* tinker, tailor, soldier, spy.

Speed destroyed the Summer of Love. It brought the countercultural revolution to its premature close. While speed in tablet form was common and readily available in the 1950s, by the 1960s, a new generation of users began shooting it intravenously. These were the speed freaks. And when the speed freaks came into contact with the acidheads, their worlds collided. Speed, as one journalist put it, "had the distinction of being one of the very few drugs stigmatized within a drug culture of seemingly limitless tolerance."[22] Allen Ginsberg tells us that "speed is anti-social, paranoid making. All the nice gentle dope fiends are getting screwed up by the real horror monster Frankenstein Speedfreaks who are going around stealing and bad mouthing everybody."[23] Acidheads try to build sustainable, non-violent, non-fascist communities; speed freaks are inherently destructive of community. If a bunch of speed freaks rent a house together, then within a month the police are going to be there responding to a fight with a broken bottle, a fractured skull.

Speed is America itself. This was a trope that theorists and journalists and sociologists of the 1960s couldn't resist making over and over. Speed mimics not only the hustle and bustle of the American consumer, but the unthinkable violence of Vietnam. Speed is madness; speed is America; speed is fascism. Nobody has ever been beaten to death by a group of acidheads. Speed and acid, then, cannot be seen merely as two "drugs." They are not drugs; they are metaphysical principles. They are opposing forms of existence.

Nobody saw this more clearly than David Smith, aka "Dr. Dave," founder of the Haight-Ashbury Free Clinic. His story has been told elsewhere, but here, in short, are the details.[24] January 14, 1967, was the occasion of the first Human Be-In, a festival at Golden Gate Park's Polo Field. This is where Timothy Leary coined the phrase "turn on, tune in, drop out," an expression of the assertive refusal to participate. After that, the word on the street was that over 100,000 young people would descend on Haight-Ashbury that summer. Dr. Dave and his colleagues at the University of California Medical School anticipated the medical problems that these young people would experience, either from drug overdoses or from living in cramped and unhygienic conditions. So, they opened the Haight-Ashbury Free Clinic, with the motto that it would provide "medical treatment free from red tape, free from value judgements, free from eligibility requirements, emotional hassles, frozen medical protocol, moralizing, and mystification." The public hospitals would have likely seen a drug overdose as a criminal problem and thereby alerted the police.

One of Dr. Dave's roles in the clinic was to track changing patterns of drug use in the Haight, and to that end, he worked with criminologist Roger Smith, otherwise known as the "Friendly Fed," to compile data. What they noticed over the period from 1967 to 1969 was that the patterns of drug use were shifting rapidly from acid to speed. The same pattern was seen in other hubs like the East Village and the Sunset Strip.

As a result of the confrontation between the acidheads and the speed freaks, Dr. Dave tells us, many of the acidheads, at least those with means, fled to smaller communities where they could live in a manner consistent with their principles. But what is more interesting about this report is that Dr. Dave grasped the metaphysical significance of this confrontation. It is *inevitable*, he thought, that the speed freaks would drive out the acidheads; it couldn't *not* happen that way:

> Because of the violent characteristics of the [speed freak], the hippies have moved to the country where they can establish small rural communes which tolerate and reinforce their belief systems. Urban areas such as the Haight-Ashbury can never be a permanent haven for the acid subculture, because in the conflict of *speed freaks* vs. *acid heads*, *speed* always drives out *acid*—as in the broader society the philosophy of violence dominates the higher aspirations of nonviolence, peace and love.[25]

Though speed brought the Summer of Love to its premature conclusion, it also left a lasting impact on psychiatry itself: acid could no longer be thought of as a psychotomimetic drug. It was now conceived, *even inside of "medical psychiatry,"* not as a mad-making drug, but as a sane-making drug.

IV.

Somebody on massive quantities of speed can write and talk in ways that seem nonsensical; consider Sartre's *Critique of Dialectical Reason* of 1960, produced under large amounts of Corydrane.[26] But this is not yet madness. For if persuaded to make the effort to explain their reasoning, speed users can generally do so. The speed user's reasoning is amplified; the velocity of his reasoning is increased, he speaks in abbreviations and ellipses, but his thought is not disordered. The absence of thought disorder *proper* was one reason that researchers in the early 1960s rejected the use of speed as a psychotomimetic drug. Nonetheless, by the late 1960s, psychiatrists began to reassess the use of LSD as a model of schizophrenia and the competing usages of acid and speed as psychotomimetic drugs. There was an emerging consensus that *speed is a better psychotomimetic than acid*. Incredibly enough, the psychiatrists who promoted this consensus utilized the language and conceptualizations of the American counterculture in building their case.

First, consider Angrist and Gershon, who worked at Bellevue Hospital in the late 1960s and also witnessed the transition from LSD overdoses to speed overdoses. They noticed that many of the people who came in would have been diagnosed as having a psychotic episode if not for the knowledge that they were overdosing on speed; speed overdoses could be indistinguishable from psychotic episodes. So, they began exploring the thesis that speed could make for a powerful psychotomimetic drug. As part of their reasoning to this end, they argued that LSD simply could not play this role because of the differing characteristics of the speed freak and the acidhead. *Acidheads aren't crazy; they are gripped by a philosophical or religious worldview but they are not mad*: "Because of . . . their sociopathy and their frankly hedonistic reasons for drug use, [the amphetamine users] resemble heroin addicts as a group far more than the philosophically and religiously preoccupied and less sociopathic hallucinogen users."[27] Even in a psychiatric wing of a large public hospital, we are seeing this moral evaluation, this moral differentiation, of the acidhead and the speed freak; they are clearly recognizing that the use of acid is part of a philosophical worldview with an ultimately noble end and that the use of speed is not.

Two years later, in response to their work, the American neuroscientist Solomon Snyder, founder of the dopamine hypothesis of schizophrenia, also argued forcefully against the claim that LSD should be considered a model schizophrenia, and he used a similar, morally tinged, distinction. LSD, he wrote, in an unmistakable allusion to Huxley, throws open the doors of perception: "The mental state elicited by psychedelic drugs is one of greatly enhanced perception of oneself and one's environment. Similar states occur during mystical and religious introspection and when an individual is profoundly moved by emotions or external events."[28]

In addition to displacing LSD as psychiatry's psychotomimetic, Snyder, like Angrist and Gershon, argued that speed should take its place. But what of the problem of thought disorder, and its supposed absence in speed? In the early 1960s, researchers reasoned that speed could *not* be a model schizophrenia because it did not mimic that symptom in its entirety. Researchers responded to that apparent disanalogy in two ways. Angrist and Gershon, through their observation of patients who had overdosed, as well as through their own controlled experiments with speed, insisted that it did, in fact, mimic thought disorder. In one particular study, four subjects volunteered to take speed so that the researchers could observe their behavior; of those four, one young man exhibited symptoms that resembled thought disorder. He began writing feverishly and launched into an

> agitated philosophical diatribe with riddles that made little sense. For example, "one man goes to school, the other can't. Then the other 'cuts out' say, 'fuck you, buddy.'" This he explained meant that there is no brotherhood in the world. Questions the meaning of gold and source of its value.[29]

Later the same subject started referring to himself excitedly as a prophet. Angrist and Gershon took that as proof that *speed could induce thought disorder in otherwise normal people*:

> These phenomenologic features [auditory hallucinations and thought disorder] give amphetamine psychosis a greater resemblance to naturally occurring schizophrenia than the states induced by other psychotomimetics. . . . This clinical resemblance of amphetamine psychosis to schizophrenia justifies study of its mechanisms of pathogenesis.[30]

Snyder developed a very different explanation for the supposed absence of thought disorder in speed users. His explanation was that speed probably has two different chemical components that act on two somewhat competing neurotransmitter systems. One component acts on the dopamine system, and if that component acted alone, then it *would* yield a clinical condition that would be indistinguishable from schizophrenia. However, speed, as it were by accident, has another chemical compound that probably acts on another transmitter system, the norepinephrine system, and when it acts on the norepinephrine system, that tends to dampen the effect of the first.

For Snyder, this is the inner life of the speed user: *first*, the dopamine system is activated, which alone *would* make you a raving lunatic. But, *second*, it acts on the norepinephrine system, which stifles the incipient madness. Phenomenologically, what it feels like is this: one takes speed, and begins to feel a kind of inner confusion, the paradigmatic thought disorder of schizophrenia. But then—and this is where the norepinephrine kicks in—you begin to come up with a system of delusions that effectively make sense of these unusual feelings that you are having. So, the madness-making part of speed and the reason-making part of speed fuse together in such a way as to yield an interlocking system of delusions. But this system of delusions is not itself madness, it is an *inner, compensatory response to madness*. With it, the patient "strive[s] for an intellectual framework in which to focus all the strange feelings that are coming over him as the psychosis develops."[31] When I start to become mad, say, I start hallucinating, the reasoning part of my mind desperately tries to come up with a theory to explain what is going on and to restore a sense of control. So, for example, if I begin hearing voices in my head, the reasoning part of my brain suggests that, most likely, the last time I visited my dentist, the dentist implanted a radio transmitter in my tooth and I am hearing the buzz of that. And while, in a sense, I am a "madman"—who else but a madman forms the idea that his dentist implanted a radio device in his tooth?—the delusions stave off an even deeper madness. Delusions are the only thing that enable me to continue to navigate my environment; the delusions rescue me from madness.

Amphetamine psychosis, and in particular, the idea of speed as a psychotomimetic drug, played a crucial role in the transformation of American psychiatry in the 1970s. It was one of the pillars of the dopamine hypothesis of schizophrenia, the idea that schizophrenia stemmed from the overproduction of dopamine in the brain. The dopamine hypothesis, in turn, transformed psychiatry by allegedly demonstrating that a major mental disorder could be explained entirely by the disruption of a single neurotransmitter system. The dopamine hypothesis, then, laid the foundation for the so-called second biological revolution in American psychiatry. In that respect, it helped to pave the way for that symbol of contemporary psychiatry, the momentous third edition of the *Diagnostic and Statistical Manual of Mental Disorders*. For our own purposes, the function of the *DSM-III* was to expunge reference, systematically, to teleology from American psychiatry. There is no longer the good madness and the bad madness, the revolutionary madness and the madness of the disheveled man drooling on his beard. There is only one kind of madness, and that is the madness that stems from something going wrong inside of you. There is only the madness that happens when a faculty of the mind breaks down. There is only Kantian madness. The chief task of the nosologist in this particular sociohistorical era is to document rigorously, to catalog, to systematize, all of the diverse ways that the mind can fail.

14

Confronting the Wounded Animal

The massive third edition of the *Diagnostic and Statistical Manual of Mental Disorders* (*DSM-III*) of 1980 marked a revolution in American psychiatric thought. But what, precisely, is it attempting to do? Many historians and philosophers have written on the *DSM-III* and the intellectual changes that it wrought for the project of psychiatric nosology, but there is an important aspect of its character that has largely gone unnoticed. The purpose of the *DSM-III* is to complete a task that the *DSM-II* left unfinished: to strip teleology from psychiatric classification (Section I). In its place, it insists that *all* disorders stem from "behavioral, psychological, or biological *dysfunctions*" (Section II). Glancing back at the history of the *DSM-III* definition of "mental disorder" can help us understand the contours of the concept of dysfunction invoked there (Section III). Once we look at that history, we can appreciate a problem that many psychiatrists, in the lead-up to the publication of *DSM-III*, openly struggled with: what is a dysfunction? What is it, precisely, for something inside of you to *go wrong*? And whence the norm by which we make that assessment? Here, we consider several attempts on the part of psychiatrists to use evolutionary theory to explain this critical idea. Kendell, and later Spitzer and Endicott, argued that all mental illnesses must involve an "intrinsic biological disadvantage," to be measured in terms of increased mortality or reduced fecundity (Section IV). Donald Klein, in contrast, argued that mental disorders must involve a failure of *evolved design* (Section V). Klein's attempt raises a problem: why would one think that evolutionary reasoning would tend to *justify*, rather than *subvert*, our existing system of classification? This problem is taken up in Chapter 15. Finally, we must pause to ask whether the contemporary Research Domain Criteria (RDoC) project represents a fundamentally new orientation for nosology. We argue that it does not (Section VI).

Madness. Justin Garson, Oxford University Press. © Oxford University Press 2022.
DOI: 10.1093/oso/9780197613832.003.0015

I.

Much has been written on the transformation of psychiatric nosology in the second half of the twentieth century. But what is the essential character of this transformation? Its essential character is this: *dysteleology replaces teleology*. In 1952, all forms of madness represent so many contrivances, so many ingenious, albeit unconscious, strategies that individuals deploy to negotiate their inner and outer worlds. In the second half of the twentieth century, the forms of madness become, instead, just so many different ways of going wrong, so many different ways that things can err inside of you. The transitions within psychiatric nosology, beginning with the *DSM-II* of 1968, continuing with the *DSM-III* of 1980 and its multiple and voluminous revisions, and coming to fruition with the RDoC system of today, represent the progressive triumph of Kantianism.

In order to appreciate the character of this transformation we must expose three historiographical errors about nosology. The first is that one of the chief characteristics of the *DSM-III* is that it replaces a psychodynamic, Freudian view with a medical, non-Freudian one. This is not quite right: the urge, the impetus, to dismantle the Freudian system of thought already began with the *DSM-II*. The *DSM-III* did not initiate the process of eliminating reference to psychodynamic ideas, and, more generally, to teleology, but completed it. It has the character of a mop-up job.

The second error is that the *DSM-III* represents something like a biological takeover of the profession; with it, the biological psychiatrists hoist their flag over psychiatry just as the American soldiers did over Iwo Jima in 1942. The *DSM-III* did not represent the replacement of a psychogenic point of view with a biogenic point of view; it represented the culmination, the end point, of the replacement of a teleologically centered worldview with a dysfunction-centered view. The transformation, the notion of dysfunction at play in the *DSM-III*, is neutral between biogenic and psychogenic: there can be biological dysfunctions, psychological dysfunctions, behavioral dysfunctions. It was not an attempt to put psychiatry in a biological straitjacket. (Historically, that would not make much sense, given that many of the *DSM-III* Task Force members were not particularly biologically oriented.)[1]

The third error is that the Research Domain Criteria (RDoC) project, overseen by the National Institute of Mental Health (NIMH), represents a *competitor*, a radically new principle for dividing up the space of madness. The RDoC should not be seen as a competitor but as the fruition of

the *DSM-III*'s organizing principle. While *DSM-III* and its successors announce that mental disorders must stem from inner dysfunctions, they make
no rigorous attempt to actually *discover* these dysfunctions. The purpose
of RDoC is simply to organize the way that the mad are partitioned so that
these hidden dysfunctions will become manifest. In RDoC, *the same principle—the principle of inner dysfunction—is merely deployed in a more systematic and unforgiving way*. As a matter of justice, *DSM* should give way to
RDoC; it should recognize RDoC as its legitimate heir; it should acquiesce to
its leadership. With that context in mind, we can begin to understand what
the *DSM-II* is doing.

If the purpose of the *DSM-I* was to rein in, to restrain, a kind of diagnostic anarchy at the *national* level, a goal of the *DSM-II* is to rein in this
anarchy at the *international* level. *DSM-II* announces its global intent in the
very first line of the foreword, penned by Ernest Gruenberg, head of the
APA's Committee on Nomenclature and Statistics: "This second edition of
the *Diagnostic and Statistical Manual of Mental Disorders* (*DSM-II*) reflects
the growth of the concept that *the people of all nations live in one world*."[2]
This statement is not just a bit of American arrogance. The task force of the
DSM-II had been working closely with the World Health Organization's
International Classification of Diseases (ICD); there was an authentic push
to integrate the two systems of classification. American nosology had, by the
late 1960s, become globalized.

Crucially, this globalization, this homogenization, is not just a matter of
devising a uniform nomenclature. It is global in another sense, namely, that it
proceeded from the realization that the way that categories of mental illness
are defined should not reflect one or another narrow disciplinary outlook; it
should be perfectly ecumenical. Hence Gruenberg insists upon the necessity
of expunging reference to psychodynamic concepts and verbiage from the
DSM, the need for *neutrality*:

> The Committee accepted the fact that different names for the same thing
> imply different attitudes and concepts. It has, however, tried to avoid terms
> which carry with them implications regarding either the nature of a dis-
> order or its causes and has been explicit about causal assumptions when
> they are integral to a diagnostic concept. In the case of diagnostic categories
> about which there is current controversy concerning the disorder's nature

or cause, the Committee has attempted to select terms which it thought would least bind the judgment of the user.[3]

Immediately, and with remarkable foresight, Gruenberg anticipates many of the debates that later crop up around the *DSM-III* with so much force and publicity: "Inevitably some users of this Manual will read into it some general view of the nature of mental disorders. The Committee can only aver that such interpretations are, in fact, unjustified."[4] In other words, he understands that some mental health professionals will see in the *DSM-II* the product of a kind of cabal, a secretive and insidious attempt by one particular group of practitioners to assert their own ideology over all others; they will see a cynical political ploy by the non-Freudians to gain the upper hand. The same critique was repeatedly leveled against the *DSM-III* in the 1970s; Garmezy's accusation of "territoriality" against the *DSM-III* is representative.[5] Gruenberg insists that the *DSM-II* is, rather, an attempt to purge *all* local ideologies from the system of classification. The ultimate product that issues from the task force's deliberations—the manual itself—should not be recognizably psychodynamic, or biological, or behavioral. It must be cosmopolitan and "atheoretical."

How was this global ambition actually implemented in the *DSM-II*? On the surface, the *DSM-II* is not very different from the *DSM-I*. The *DSM-I* runs to 129 pages; the *DSM-II* runs to 119. Very roughly, the two books carve up the space of mental illness according to the same general plan: there are organic and non-organic disorders; the latter divide, roughly, into psychotic, neurotic, personality, and situational disorders. A key difference, however, is that the *DSM-II* attempts to minimize, to the extent possible, purpose and end. The first major change is the elimination of the term "reaction" from many of the diagnostic labels. Gruenberg himself gives the removal of "reaction" as a specific example in which the task force expunged a term because it was too weighed down with disciplinary connotations. We will pick up Gruenberg where we left off:

Inevitably some users of this Manual will read into it some general view of the nature of mental disorders. The Committee can only aver that such interpretations are, in fact, unjustified. Consider, for example, the mental disorder labeled in this Manual as "schizophrenia," which, in the first edition, was labeled "schizophrenic reaction." The change of label has not changed the nature of the disorder, nor will it discourage continuing debate about its nature or causes.[6]

What was so problematic about "reaction"? As noted earlier, "reaction," even more than "neurosis"—which at any rate only stands for one subdomain of mental disorder—carries within it a rich teleological perspective. This notion of a reaction, as in "schizophrenic reaction," originates with Adolf Meyer, for whom all so-called pathological deviations of mental life are "reactions" to crisis situations that the organism faces:

> The concept of substitutive reactions is meant to keep us from wandering from the ground of the experimental formula of investigation. . . . To realize that such a reaction is a *faulty response* or *substitution of an insufficient or protective or evasive or mutilated attempt at adjustment* opens ways of inquiry in the direction of modifiable determining factors.[7]

Years after the publication of the *DSM-II*, Theodore Millon, a member of the *DSM-III* Task Force, noted in retrospect that this teleological purge did not go far enough:

> Although the principle of eschewing theory-based positions was both explicit and authentic, it does appear curious, to say the least, that the only major shift of this character was the consistent elimination in the *DSM-II* of the syndromal adjective "reaction" from a diverse group of diagnostic labels. . . . Untouched, however, were equally doctrinaire concepts, such as neuroses or psychophysiologic disorders, which derive their logic from "psychoanalytic" formulations.[8]

As a rule, the only place that the language of "reactions" is maintained is in the adjustment disorders ("transient situational disturbances"), which are defined by their teleological orientation or goal-directedness, their being "acute reaction[s] to overwhelming environmental stress."[9] These, too, were retained in the *DSM-III*.

The second major change, in addition to the removal of "reaction," is the wholesale elimination of these teleological characterizations of the three main types of non-organic mental disorders—psychotic, neurotic, personality— as three different ways of coping with inner and outer stressors. The entire section in the *DSM-I* that introduced these major categories and announced their basic Freudian orientation has, in the *DSM-II*, disappeared: the book jumps right into its subdivisions without any preliminary overview. True, some teleological language remains. The whole category of the neuroses remains; the authors of the *DSM-II* note, in passing, that neuroses can be seen as a transformation of anxiety, which "may be felt and expressed directly, or it may be controlled unconsciously and automatically by conversion, displacement and various other psychological mechanisms."[10] It is also true that in the single-paragraph overview of schizophrenia, there is passing mention of the possibility that hallucinations and delusions "frequently appear psychologically self-protective."[11] But it is more accurate to see these isolated statements as *vestigial*—with all of the danger that interpretive strategy entails.

Having understood what the *DSM-II* is up to, we can see the *DSM-III* in its true light. This major transition that historians note, the fact that it finally eliminated the term "neuroses," represents more of a *mopping up* of the task that the *DSM-II* failed to complete. The *DSM-II*'s mission was to eliminate such language, but it failed to do so, probably because of the centrality of the psychosis/neurosis distinction and its psychodynamic interpretation. Not only does the *DSM-III* eliminate neuroses, but it fragments the unity of the category. Instead of one category that encompasses anxiety disorders, paraphilias, and so on, each has a separate category. So, it is not a question of keeping the category and changing the term, but *abolishing the neuroses as a natural kind*.

II.

The *DSM-II*, in eliminating reference to teleology—at least as much as was possible at the time—created an intellectual vacuum. What is its organizing principle? For every classification is a classification of *something*. Another way to put the question: what is a mental disorder? What exactly is being classified in this book? In 1968, nobody felt an acute need to say what mental disorders were, to fix the essence of madness and deduce, from that essence, a principle of inclusion and exclusion, as Goldstein sought to do with physiology as a whole. The *DSM-III* fills this void in a specific way: we are, here, classifying *the ways the mind can go wrong*. The book is a compendium of the ways the human mind can err.

The activity of classifying requires an organizing principle, and it requires that this principle be somewhat explicit in the minds of those who are in the

business of classifying. But to state the organizing principle of one's classification is simply to state, more or less explicitly, what the classification is a
classification *of*. If I want to classify the world's ethnic groups, I must start
with a more or less explicit characterization of what ethnicity is; this gives me
a principle for which groups to include and which to exclude. This is where
scientists turn to "philosophical" considerations. They find ways to signal or
to indicate that they are "taking off" the scientist's hat and "putting on" the
philosopher's hat. In the 1970s, in the lead-up to the *DSM-III*, we see an explosion of self-conscious attempts on the part of psychiatrists to engage in
"philosophical" deliberations about the nature of mental illness. We must
turn to the *DSM-III* definition to get a preliminary fix on this new organizing
principle, to trace its boundary; this will allow us to work backward through
the passage's earlier drafts so as to continue to refine our preliminary outline:

> In *DSM-III* each of the mental disorders is conceptualized as a clinically
> significant behavioral or psychological syndrome or pattern that occurs in
> an individual and that is typically associated with either a painful symptom
> (distress) or impairment in one or more important areas of functioning
> (disability). In addition, there is an inference that there is a behavioral, psy-
> chological, or biological dysfunction, and that the disturbance is not only in
> the relationship between the individual and society. (When the disturbance
> is *limited* to a conflict between the individual and society, this may repre-
> sent social deviance, which may or may not be commendable, but is not by
> itself a mental disorder.)[12]

In the *DSM-III-R* (the revised edition of 1987), this characterization is
amended slightly. First, puzzlingly, it adds that this behavioral, psychological, or biological dysfunction must be "in the person."[13] (*Where else would
it be?*) Second, it explicitly excludes from the extension of "mental disorder" those conditions that can be seen as *reactions to stressful events*: the
syndrome must neither be "merely an expectable response to a particular
event, e.g., the death of a loved one," nor deviant behavior, "e.g., political,
religious, or sexual."[14] What we need to appreciate is that its reference to
a "behavioral, psychological, or biological dysfunction" is not a slip; it
is not a last-minute inclusion; it is the result of multiple discussions and
confrontations; it is, strictly speaking, an intervention. But before we turn
to the history, we must ask about some very general structural features of
this definition.

The first thing to note is that the notion of function or functioning is *doubled* in the passage. First, a mental disorder is associated with, among other things, "impairment in one or more important *areas of functioning.*" Shortly thereafter, we are told that the syndrome must be a manifestation of a dysfunction *in the person*: a "behavioral, psychological, or biological *dysfunction.*" So there are *two* sorts of failures of function at play, not one: to have a genuine mental disorder you must fail, as it were, twice over. The *inner* failure of function causes the *manifest* syndrome, which, in turn, causes the failure in an *area of functioning.* A neurotransmitter abnormality (the inner failure of function) causes hallucinations and delusions (the syndrome), and the delusions cause me to miss work or school (failure in an area of functioning).

What, then, is an "area of functioning"? These areas are, first and foremost, *realms of social life*; they concern one's "place" in the larger scheme of society. Fortunately, in the *DSM-III*, many of the descriptions for specific disorders identify, as part of their very "inclusion criteria," the area of functioning that has been wanting: a criterion for schizophrenic disorder is "deterioration in functioning in such areas as work, social relations, and self-care."[15] Attention deficit disorder with hyperactivity is associated with impairment in "academic functioning"; avoidant disorder with "social functioning in peer relationships."[16] One is, in the first place, a student, a teacher, a doctor, a lover, a friend, with all of the rights and responsibilities such a role entails.

But what this working definition tells us is that *simply* the lapse in one or more "areas of function" is not constitutive of having a mental disorder. If Johnny fails to bathe for a week, if he shows up late to work repeatedly, if he drops out of school, these might not reflect disorders. Perhaps he has acquired a "philosophy"; perhaps he is smoking pot and reading Timothy Leary, and his behavior represents an explicitly formulated wish not to be society's pawn. That, alone, is a mild social deviance, "which may or may not be commendable," but it is not disordered since, by hypothesis, it does not yet stem from an inner dysfunction. The second failure has not yet taken place and he is not mad. For this to be madness, these outward, observable behaviors must be an expression of a dysfunction *inside* of him. The core problem, then, is this: what is this dysfunction inside the person, this dysfunction that is necessary for a syndrome to constitute a disorder? What are the features of this dysfunction? What are its empirical hallmarks? We do not wish to have, as Kendell accuses the *DSM-III* Task Force of having, a definition that is "vaguely worded [enough] to allow any term with medical connotations to be either included or excluded in conformity with contemporary medical opinion."[17]

Return to the second part of the *DSM*'s working definition: "There is an inference that there is a behavioral, psychological, or biological dysfunction, and that *the disturbance is not only in the relationship between the individual and society.*" The *DSM-III-R* clarifies that deviant behavior alone, "e.g., political, religious, or sexual," does not alone implicate dysfunction, and that the dysfunction must be "in the person." Protesting a repressive government, or flaunting social mores, or satisfying unusual sexual preferences does not yet constitute a disorder. So, what is an *inner dysfunction*?

These passages allude to three different features that the relevant concept of dysfunction must possess. The first feature of this inner dysfunction is its *intrinsicality*. It is something "in the person" that *causes* you to flaunt social mores or enter into conflict with society. It is something intrinsic to you that has causal power. But not just *any* inner condition that causes conflict with society is a mental disorder. Being a racist can cause you to come into conflict with society's norms but, as Spitzer insisted, it is not a disorder and is not generally held to be one today.

A second characteristic is its *domain neutrality*. This dysfunction, whatever it amounts to, need not be a *biological* dysfunction. It can be a "behavioral, psychological, *or* biological" dysfunction. One thing this remark helps to show, in addition to many other things, is that the *DSM-III* Task Force did not think of their primary mission in terms of a biological takeover of the field. Whatever this dysfunction turns out to be, they insisted on its domain neutrality. An appeal to an inner dysfunction should *not* be a matter of planting a biological flag over psychiatry. So, while the notion of dysfunction is intended to be substantive and non-trivial, while it is intended to carve out a special space for the psychiatrist, to delineate her sphere of professional jurisdiction, it is not intended to privilege one psychiatric orientation over another (biological, psychodynamic, etc.).

This brings us to the third characteristic, which is *dysteleology*. This inner condition has to represent a *way of going wrong*. But—and this is absolutely pivotal, and recognized by all parties to the discussion—to the extent that we wish to characterize this inner state as a "way of going wrong," the standard for going right must be rendered explicit *and it cannot be a standard derived from the social sphere*. For then, the insistence that there be a dysfunction, *in addition to* "a conflict between the individual and society," would not amplify the definition. If a mental disorder is not merely "a conflict between the individual and society," but it represents the outcome of an inner state, and this inner state is a way of going wrong, then the standard by which we judge the inner state to be a "way of going wrong" cannot itself be drawn from our social norms, or else we would not have added anything to the original definition; the statement ("in addition, there must be an inner dysfunction") would not be circular; it would be redundant. Any behavior that society detests strongly enough would become, by that definition, disordered.

We can at last approach the fundamental puzzle that all of the psychiatrists involved in constructing a definition of mental disorder are trying fervently to solve: how to characterize what it is for *something to go wrong inside of you* without appealing to any standards or norms drawn from the social sphere? As Donald Klein, one of the interlocutors in this debate, put it, "Can one define disease in a fashion conceptually independent of illness?"[18] For Klein, "illness" is *partly* a social category; it involves the failure to perform one's social role adequately; here, he wants to know if we can define disease independently of that, independently of *all* reference to society's standards and expectations. Put differently—and we will return to the issue at length below—the question is one of explicating, rigorously, *what it is for something to go wrong inside of you.* That is, *what is it* for something inside of you to go wrong, not because people think it's wrong, but because of an intrinsic wrongness about it? Until we can say what that is, the scientific status of psychiatry remains problematic.

III.

The *DSM-III*'s working definition of disorder did not arise in a vacuum. It has a special history; it went through several rounds of revisions. In March 1973, Walter Barton, then chief executive of the APA, sent a memo to the chairman of the APA Council on Research and Development informing him of the board's instruction to appoint a task force to revise the *DSM-II* and prepare the *DSM-III*. In addition to recommending a more problem-oriented and quantitative diagnostic system, the memo also recommends "the formation of a Task Force to Define Mental Illness and What Is a Psychiatrist," and that this definition be used as a preamble to the *DSM-III*.[19]

The very fact that there is a perceived need to "define mental illness and what is a psychiatrist," the fact that we need this kind of philosophical prolegomenon, points to a new kind of consciousness emerging among members

of the APA: it implies that their position has been challenged; they feel themselves under an existential threat. If somebody asks me, "What is a philosopher?," the kind of answer that I will give depends crucially on the context of
the question. If the person asking the question is a casual acquaintance, then
it would be enough to say what a philosopher is by ostensive definition: *we* are
the ones who deal with free will, altruism, mind and body, right and wrong,
God, knowledge, etc. If, however, the question is coming from the provost of
a university who is considering eliminating the philosophy program to save
money, then I will give a very different kind of answer: I will say something
about how science creates conceptual puzzles that science, by its very nature,
cannot answer, but which it must answer in order to make progress, and that
fruitful dialogue between philosophy and science will unleash the potential
for this extraordinary progress, and then we will have flying cars and end
world hunger, etc. If, instead, the question is coming from a psychologist
who is of the opinion that the psychologists are already doing everything the
philosophers are doing, so philosophy is redundant, then I will give a somewhat different answer as well. And in *this* context, in 1973, to be asked to
form a committee to "define mental illness and what is a psychiatrist" betrays
a consciousness of a challenge, a threat; the committee is formed under a
certain amount of duress; this is an antagonistic arena: I must explain myself,
defend myself, justify my existence to a hostile "outside."

Who are these challengers? Who, exactly, are the people who are
questioning the scientific status of psychiatry? Though historians and
philosophers have covered this area extensively, there are two main groups
who seek to challenge the authority of psychiatrists.[20] On the one hand, we
have people outside of the mental health profession, the anti-psychiatrists,
sociologists, intellectuals, artists, filmmakers, who are questioning the *authority* of psychiatry. This is a demand that psychiatry give an account of
the basis of its authority to classify people as mentally well or sick. It is a
demand for a certification—a particularly pressing issue in 1973, the year
that the APA eliminated "homosexuality" from the *DSM-II* in response to
activism. This demand for certification, for an account of the source of its
authority, goes hand in hand with a culture of anti-psychiatry that repeatedly
challenges the very distinction between madness and political resistance.
The idea of *madness/acid* and the idea of *fascism/speed* are generalized and
expanded, they take on metaphysical dimensions as light to darkness, good
to evil, peace to war, and along with that—and apparently paradoxically—
the idea of two quite distinct forms of madness, a good madness and a bad

madness: a good madness, which is a form of resistance against fascism, and a bad madness, which takes place when the journey of self-discovery that constitutes the good madness is thwarted or crushed, and the thwarting of which is itself due to fascist social and political forces. Fascism cannot tolerate madness, in principle, but it can tolerate a heavily medicated, babbling lunatic, quarantined on the back wards of an asylum. So, fascism is necessarily "on a mission" to transform good madness into bad madness. And so, this certification, this demonstration of the source of the psychiatrist's authority, *must also be* a demonstration that the psychiatrist is not herself a part of, a working component of, the fascist machine that transforms good madness into bad madness.

And the second fight, the second contest, takes place among mental health professionals, with psychodynamically oriented psychiatrists who are bitter about what they see as the systematic elimination of their perspective from official APA publications, and with psychologists who believe themselves to be competent to resolve at least certain forms of mental illness without a medical degree. These are the psychiatrists and psychologists of whom Gruenberg warned as early as 1968, who would inevitably see in the *DSM-II*'s push toward an atheoretical language an insidious takeover of the profession by biologically oriented practitioners. So, this certification of authority to distinguish the mad from the sane must, at the same time, reaffirm the medical orientation of psychiatry. Put differently, when I inform you—when I provide for you—an account of the source of my authority to distinguish the mad from the sane, it must also be apparent to you, without my needing to say anything else, that madness, or at least certain of its forms, is the kind of thing that demands attention by a person with a medical degree.

One way I could solve both demands at once—the demand to legitimate my authority to distinguish the sane and the mad, on the one hand, and the demand to validate the appropriateness of the medical perspective, on the other—is to convince my interlocutor that *it is precisely the fact that I am a medical professional that gives me the authority to distinguish between mad and sane.* In other words, it is not as if I should begin this *apology* for the psychiatrist along the following lines: "For such and such reasons, I have authority to distinguish between madness and sanity, and also, I happen to be a medical professional." Rather, these two statements must relate as antecedent to consequent: it is because I am a medical professional that I have authority to distinguish madness and sanity. Mental disorders must be diseases.

Now we can understand, perfectly and fully, the appeal to dysfunction in the *DSM-III*'s definition. The purpose of the appeal to "dysfunction" is not to affirm a *biological* rather than a *psychological* or *behavioral* perspective; that would completely miss the "functional role" of the concept. It is to affirm that mental disorders are diseases, and *hence* that the physician has the special authority to delineate and treat them.

The role of the concept of dysfunction as the logical link between mental disorder and disease becomes apparent in the earlier drafts of this passage. In 1973, Spitzer was appointed to chair the *DSM-III* Task Force. The Task Force to Define Mental Illness and What Is a Psychiatrist never materialized. In lieu of that committee, Spitzer sent out an open call for papers on the topic of what is a mental disorder. Spitzer and Jean Endicott offered the first essay; Donald Klein offered another. The version that ultimately appeared in the *DSM-III*, perhaps unsurprisingly, is a minor revision to the Spitzer-Endicott definition.

In 1976, Spitzer and Endicott gave a presentation to the APA; material from that presentation was incorporated into two follow-up papers, published in 1977 and 1978, respectively. In the 1977 paper, a short preamble precedes the definition:

> Efforts on the part of some members of the Task Force to arrive at a comprehensive definition of mental disorder have been deemed generally less than satisfactory. However, these efforts have made explicit some of the principles that have been applied in determining which conditions are included in *DSM-III* as mental disorders, and which are variants of human behavior outside the direct professional responsibility of the psychiatric profession. These principles help to avoid an overly broad definition of mental disorders that would view all individual and social unrest or problems of living as psychiatric illness, and at the same time justify the designation of mental disorders as a subset of medical disorders.[21]

This passage both affirms the authority of medical professionals to determine which conditions are mental disorders and which are mere "problems of living" (invoking Szasz' famous phrase). What immediately follows is an early version of the *DSM-III* definition:

> [Mental disorders,] in their extreme or fully developed form . . . are directly associated with either distress, disability, or, in the absence of either

of these, disadvantage in coping with unavoidable aspects of the environment. Furthermore, they are not quickly ameliorated by simple nontechnical environmental maneuvers or informative procedures and do not have widespread social support. Because of these features, there is an implicit assumption that *something is wrong with the human organism* and there is a call to the profession to develop and offer preventative or therapeutic measures.[22]

Crucially, the difference between this passage and the *DSM-III* passage reveals, clearly, the impetus behind the latter definition. The first part is fairly similar: the phrase "distress, disability, or . . . disadvantage" points to a failure in a socially defined "area of functioning." However, instead of saying that there must be a dysfunction inside the person, they say that there is an "implicit assumption that *something is wrong with the human organism*." Saying that there is a "dysfunction," then, is another way of saying that something is "wrong with the human organism," and the *way* in which something has gone wrong, the *way* in which this mysterious standard has been violated, calls for medical intervention. In the pages that follow, they tell us that they endorse what they call the "medical model," that is, "the working hypothesis that there are organismic dysfunctions which are relatively distinct with regard to clinical features, etiology, and course. No assumption is made regarding the primacy of biological over social or environmental etiological factors."[23]

We have finally discovered the key to unlocking the secret of mental illness: when we confront the mad patient, we do not confront, in the first place, a citizen of society. We do not confront a person, or even a human being; we confront the organism. Moreover, we do not confront an organism in the joyful process of self-actualization. We confront a *wounded* organism. In the final analysis, the *rule* by which we judge that something has gone wrong is not drawn from society. It is drawn from life itself. Life itself furnishes us with the rule by which to judge that things are going right or things are going wrong.

We will look at three attempts that psychiatrists made in the 1970s to articulate what dysfunctions are: R. E. Kendell's "biological disadvantage" definition, Spitzer and Endicott's "operational" definition, and Klein's evolutionary definition. The latter two attempts take their intellectual inspiration from the first: both of the latter papers are framed, in part, as responses to the inadequacies of Kendell's attempt. So, it is to Kendell that we will now turn.

IV.

A dialogue on what mental disorders are must be *philosophical* in nature. It must be a kind of discourse that prizes *truth* above all narrow self-interest. Put differently: we all have different interests, given our professional specializations, our social identities, our stations in life: I am a psychiatrist; you are a psychologist; he is an angry young man who has been reading Thomas Szasz and who is skeptical about the value of our disciplines. But when we think about what mental disorders are, it is important to feel confident that the definition I give does not merely restate, in abstract terms, my own professional biases. When we engage in this conversation—for that is what it must be, a conversation—about the meaning of mental disorder, we must come at the topic not as a psychiatrist, a psychologist, an anti-psychiatrist, but simply as thinkers, as *people*, who share a common concern, namely, the concern to promote psychological well-being, and who are earnest and competent seekers of truth. The definition of "mental disorder" that we arrive at must, therefore, be one that *any intelligent person would arrive at through earnest reflection*, regardless of their station in life or their personal interests or their stake in the outcome. It is only *after* we have achieved an ideologically neutral definition that we can begin to resolve, in a rational manner, the clash of opinions between psychiatrists, psychologists, and anti-psychiatrists: "before we can begin to decide whether mental illnesses are legitimately so called we first have to agree on an adequate definition of illness; to decide if you like what is the defining characteristic or the hallmark of disease."[24]

At the outset of his paper, Kendell announces the philosophical—or as he puts it, the "logical"—character of his inquiry. When *physicians* think about what diseases are, they, being by nature practical and opportunist, simply begin by trying to discover individual diseases and how to cure them. When one asks the physician, *what is disease?*, one is merely offered an open-ended list of known diseases. The philosopher/logician, however, begins the other way around. The logician begins by defining what he means by "disease" as a whole, and then "produce[s] individual diseases by sub-dividing the territory whose boundaries he had thus defined."[25] It is, of course, a historical accident, one born of practical need, that as doctors, we have started the other way around. So, *in this discourse*, we will take off the physician's hat, and put on the hat of the "logician." And this is what the logician does: she first *empties* her mind of prejudice, interest, and such, and then she *surveys* the different disease concepts that have been proposed over time, and then *inquires* as to whether the membership conditions entailed by that disease concept are necessary and sufficient for capturing most or all of what we pre-theoretically take to be diseases. If so, then we have completed our search; if not, we must continue. In principle, *anyone* who follows this procedure, earnestly and competently, will arrive at the same conclusion.

Kendell, then, proceeds to reveal the inadequacies of the traditional disease concepts. Must diseases be associated with a demonstrable physical lesion? They are not always so associated: hypertension, for example, is a disease but it is not associated with a lesion. Is being a disease simply being statistically rare? Clearly, that is not quite right either: we cannot *identify* disease with the statistically atypical; in that case, being an Einstein or a Curie would be a disease. Borrowing from Scadding, however, he arrives at a crucial hint: a disease inherently places its bearer at a "biological disadvantage." Though Scadding did not elaborate, presumably "biological disadvantage" entails either an increase in mortality or a reduction in fecundity, the currency that the evolutionary biologist trades in. On the "biological disadvantage" criterion, diabetes is a disease because it increases mortality; ovarian cysts because they reduce fertility. This criterion sounds promising indeed.

With the "biological disadvantage" criterion, we invoke the language and conceptualizations of evolutionary biology: very simply put, we are invoking the notion of *fitness*, as in "survival of the fittest." What makes something a disease, at first pass, is that it entails a reduction in fitness, whether measured in terms of survival or fertility. Moreover, this definition, this "biological disadvantage" criterion, is on firm scientific footing, for the evolutionary biologist studies the origin and diversity of life *as such*, not this or that form of life; therefore, if *evolutionary biology itself* implies or inspires a certain conception of disease, then we finally have a foundation for approaching madness that transcends ideology, that transcends the social sphere entirely: *life itself* insists upon what is good and what is bad, what it is for things to be "going well" and what it is for things to be "going badly." "Going well" for a living being means surviving and proliferating; "going badly" means something obstructs or retards the impetus of life. From this eminently sensible and scientifically validated starting point we also arrive at a foolproof *procedure* for deciding, in any given case, whether or not a certain condition is a disease: we consult mortality and fertility differences between the people *with* the condition and those *without*.

How does this biological disadvantage criterion apply to the problem at hand, namely, mental illness? What we can deduce at once, Kendell tells us, is that *manic depressive illness and schizophrenia are both diseases*. That is because, on average, people with those conditions are more likely to die prematurely than those without: "there is also evidence from numerous sources that at least 15 per cent of manic-depressives die prematurely by suicide, and without treatment the mortality would be considerably higher."[26] And while he admits that the evidence of heightened mortality for schizophrenia is "less clear," we can reason in an a priori manner that "if schizophrenics were simply to be ignored and provided neither with sanctuaries where they could be fed and clothed nor with modern chemotherapy there is little doubt that comparatively few would survive to old age."[27] We can also demonstrate through our criterion that *homosexuality is a disease*: "the condition which stands out above all others in its implications for fertility is homosexuality. Although there have been few formal studies of the fertility of male or female homosexuals, it can hardly be doubted that it is drastically reduced in both."[28] The same goes for *drug addiction*, which is associated with higher mortality rates.

Immediately, however, we confront a rather obvious rebuttal to Kendell's "biological disadvantage" criterion: isn't it possible that, to the extent that certain mental disorders are associated with lowered fertility or heightened mortality, that is simply because of the social prejudice against people who have them, a prejudice that psychiatry as an institution fosters? "Scheff and other sociologists would argue that these handicaps may exist but are secondary consequences of the individual having been labeled as ill rather than being innate and inevitable."[29] Hence, Kendell concludes, being at a biological disadvantage cannot be *sufficient* for having a disease; there are people who are at a significant biological disadvantage but do not have diseases. As the logician puts it, biological disadvantage is necessary but not sufficient for disease. What *more*, then, what new criterion, must we add to the biological disadvantage criterion?

The additional criterion is that the disease must place the individual at what Kendell calls an *innate* biological disadvantage. Crucially, "innate" here must not be read in the sense of genetically determined, or as possessed at birth; "innate" must be read in the sense of "not depending for its negative impact on societal values and prejudices," i.e., *intrinsic*. What is it, then, for a condition to be *intrinsically* disadvantageous rather than disadvantageous because of how society treats you? It is simply this: *in a hypothetical world* in which the mad are treated no differently than the sane, would they still show a reduction in fitness? "The criterion must be, would this individual still be at a disadvantage if his fellows did not recognize his distinguishing features but treated him as they treat one another?"[30] No longer, then, is it sufficient to tabulate fecundity and mortality differences between people with and without schizophrenia; we also must engage in counterfactual reasoning: *at a nearby possible world*, as the philosophers like to say, where a person with schizophrenia is neither stigmatized nor treated with any disdain, would that person thrive just as well as someone without it?

Still, the problems for Kendell multiply. Even after we have departed from this world and cast our gaze over the merely possible worlds, we are confronted with a kind of trans-world indeterminacy: philosophers have noted repeatedly that assessing these kinds of counterfactuals depends crucially on how, exactly, we imagine these alternate worlds to be. Are we to imagine a world in which people with schizophrenia are cherished, in which they are thought to have special gifts? Is this a world in which people with schizophrenia are considered to be closer to God, to be prophets and shamans? Or is it a world in which the concept of schizophrenia simply doesn't exist? Is this a world in which forms of psychological difference simply sink into oblivion, in which they have no more significance than having freckles or a chin cleft? A deeper problem is that the very idea of an *intrinsic biological disadvantage*, if read strictly, would seem to be a contradiction in terms. For the biologist, the very concept of disadvantage is relational. A biological disadvantage is always a disadvantage *relative to what the others are doing*. Having schizophrenia is disadvantageous *relative* to those who do not have it. Being gay, if it *is* disadvantageous in this way, is disadvantageous relative to living in a straight world. There is no such thing as a condition that is intrinsically disadvantageous, disadvantageous relative to *nobody in particular*.

Kendell's inquiry has reached a dead end. Spitzer and Endicott, aware of the difficulties, take up the challenge of defining mental disorder in such a way that the link between mental illness and disease is made apparent. The key to their approach is what they call, rather loosely, "operationalization." Just like the use of "operational criteria" for defining what schizophrenia is, or panic disorder, or depression, they want to use "operational criteria" to decide whether or not something is a mental disorder *in general*. This is supposed to be something like a no-nonsense list of rules that any competent clinician would easily be able to apply to determine whether such-and-such a condition is, or is not, a mental disorder. Hence, Spitzer and Endicott first give us their definition, and then they attach, to that definition, four "operational criteria" for deciding, clinically speaking, whether the definition applies.

Still, even though they do not define "dysfunction" explicitly, they attempt to define "dysfunction" *implicitly*. In other words, we should be able to see, among the operational criteria that must be satisfied in order for something to be a mental disorder, a subset of rules that are meant to capture this somewhat elusive notion of dysfunction and, more mysteriously, the idea that something has *gone wrong* inside the human organism. They do so with their first criterion, which states that "the condition, in the fully developed or extreme form, in all environments (other than the one especially created to compensate for the condition) is directly associated with [distress, disability, or disadvantage]."[31] The key here is the parenthetical remark. In order to decide whether, say, schizophrenia is a genuine mental disorder, and not just a conflict between the individual and society, "which may or may not be commendable," *it must be associated with disadvantage in all environments but for the mental hospital.* To have a disorder is to have a condition that is universally disadvantageous; the disadvantage experienced by somebody with a mental disorder is not limited to this or that environment; it would manifest itself in every environment—but for the asylum.

Extraordinarily enough, however, in explaining what they mean by a "disadvantage," they do not rely on Kendell's notion of biological disadvantage—that is, this notion of disadvantage has nothing to do with a *fitness* disadvantage. Instead, Spitzer and Endicott simply give us a list of areas or realms of life which "are now considered, in our culture, as suggestive of some type of organismic dysfunction."[32] This list includes *impaired ability to make important environmental discriminations; lack of ability to reproduce; being cosmetically unattractive because of a deviation in kind; impairment in the ability to experience sexual pleasure in an interpersonal context.* One immediate advantage of this cumbersome definition is that, contrary to Kendell's notion of a biological disadvantage, *it does not imply that being gay is a mental disorder, because it does not impair one's ability to experience sexual pleasure in an interpersonal context.* (Since Spitzer was the one who initiated, from within, the push to delist homosexuality from the *DSM-II*, this was obviously an important desideratum.) Of course, if one changed the word "interpersonal" to "heterosexual," then being gay *would* be a disorder, according to their definition. But "heterosexual" is not the word they use. They use "interpersonal"; consequently, it is not.

An obvious problem with this list is the apparently ad hoc and culturally relative enumeration of these "areas of functioning." In other words, they are not saying: *Here is what dysfunctions are*. They are saying: *Here is a list of things that, in our culture today, are generally considered to involve organismic dysfunction: X, Y, Z.* And the failure of this attempt should be obvious: it simply does not do what it was designed to do, namely, to characterize what it is for something to be a dysfunction *irrespective* of societal attitudes, to spell out a notion of disease that is independent of illness, one that does not depend on our transient social mores.[33]

V.

Let us begin again. We will now turn to Donald Klein's definition, which was dismissed by the task force as being "overly abstruse and theoretical."[34] Its chief virtue is that it takes up, in a serious and sober way, the challenge of articulating and rigorously developing the idea that, in a mental disorder, *something has gone wrong with the organism*. He begins by situating mental disorder in the genus of the *sick role*.[35] This is an exploitative social role; there are parasitic and legitimate occupants of the sick role. When we say that a person has a mental disorder, we are saying, first of all, that that person is *entitled* to occupy the sick role; he is a legitimate occupant of that role. This is, for Klein, the core problem of defining madness: the mentally ill person must, as it were, provide us with a certificate that justifies his occupancy of the sick role.

That certificate, he thinks, is furnished by evolutionary biology: "Can we arrive at a standard [of disease] that is not simply an expression of personal preference, but is given to us by the biology of the situation? I propose that evolutionary theory allows us to infer such a standard—suboptimal functioning—and further helps us to objectively specify the optimum."[36] For Klein, disease is not, in the first place, a reduction of longevity or fecundity but rather *deviation from evolved design*. His definition appears to be *historical* and *adaptationist*: something inside of the mentally ill person cannot do what evolution designed it to do. Evolution has designed our parts and processes with specific purposes: the heart to beat, the liver to detoxify blood. Parts of our mind, too, have functions. There is something inside of us that has the function of empathy because empathy helps to promote harmonious social relationships. Disorder is simply deviation from design: "Each species has developed, through evolution, a complicated webwork of interacting adaptive, regulatory, and behavioral processes. Many of these processes are self-regulatory. Through the intricate processes of negative feedback and error comparison, these processes act as if they had goals or even a hierarchy of goals."[37]

Perhaps the most remarkable feature of Klein's definition is his confidence, bordering on faith, that *evolution will tend to justify the designation of the exemplary disorders—schizophrenia, depression, antisocial personality disorder—as design failures*. We have seen, however, from Burton, and through Freud and his followers, that it is possible to consider the various forms of madness precisely *in terms of* design, as fulfillments of design and not aberrations of it. Is it not conceivable, then, that some forms of mental illness are fulfillments of *evolved design*, too? What right do we have, what right does Klein have, in advance, to assume otherwise? He does not seem to anticipate, even vaguely, the possibility that the evolutionary criterion will fail to justify the designation of depression, schizophrenia, etc., as mental disorders. Put simply, why think evolution will underwrite our current system of mental disorder classification? The fact that it did not occur to Klein that evolutionary considerations might fail to justify our preexisting classification of madness is merely a testament to the extent to which this dysteleology perspective had, by the mid-1970s, already worked its way into psychiatric discourse. The very idea of madness-as-strategy was receding from being a live, interesting possibility, one worthy of discussion and debate, to being almost unthinkable.

Finally, in the 1990s, the very idea that one could be mad by design underwent its transition from being an *improbable hypothesis* to being a *conceptual impossibility*. This step was taken by the philosopher Jerome Wakefield— 12 years after the publication of the *DSM-III*—in his famous, albeit controversial, 1992 paper. There, Wakefield attempts to demonstrate that, *as a matter of conceptual analysis*, all disorders are harmful dysfunctions, where dysfunction is thought of, as in Klein, as deviation from evolved design.[38] My goal, here and now, is not to adopt the role of the logician: *as a piece of conceptual analysis, is being a harmful dysfunction necessary and sufficient for being a disorder?* From a historical point of view, the answer would appear to be *no*. From the time of Hippocrates, we have a lengthy tradition, or anti-tradition, of thinking of madness as by design, as purposeful, not as the failure of teleology but the fulfillment of it. Surely, therefore, it cannot be true as a matter of *conceptual analysis* that madness involves an inner dysfunction. For that would imply that nearly all of the theorists we have encountered thus far were confused, not at the level of facts, but at the level of concepts and definitions.

What is interesting for *us*, then, is not the truth of Wakefield's position, but its *audacity*. The fact that an article that announces that, *as a bit of conceptual analysis, all mental disorders are harmful dysfunctions, like all bachelors are unmarried men, or all squares have four sides*—the fact that such an analysis could appear is an indicator of a momentous historical shift. The idea that one could be *mad by design* was, at one point in history, obvious. Later, it became a hypothesis, to be seriously entertained alongside other hypotheses. Now, with Wakefield, it is a conceptual impossibility, a contradiction in terms, like a square circle. We will shortly see the irony of this: it turns out that there is little reason to believe that evolution *will* generally underwrite our current system of mental disorders. In fact, what has happened is that the infiltration of evolutionary reasoning into psychiatry, the reasoning that Klein thought would guarantee the disease status of mental disorders, has reawakened and revived the prospect that one could truly be mad by design. In that sense, the evolutionary way of thinking acts as a subversive element; it is the wedge that forces madness-as-dysfunction to make room for another way of thinking.

VI.

It behooves us to say something, even briefly, about RDoC, to demonstrate the extent to which the RDoC project is merely a flowering of the *DSM* project and *not* a fundamentally new direction for American psychiatry.[39] In 2010, Thomas Insel, then head of the NIMH, advocated for an alternative to the *DSM*, one that would focus on locating just those inner dysfunctions that the *DSM* posits. Axiomatic of this approach is that mental disorders are inner dysfunctions: "The RDoC framework conceptualizes mental illnesses as brain disorders . . . mental disorders can be addressed as disorders of brain circuits."[40]

The difference between RDoC and *DSM* is that with RDoC, we do away entirely with syndrome-based classification. Instead, we begin with a characterization of the healthy mind, the functional mind, the mind in which all of the different faculties are doing what they are supposed to be doing. These are what the framework calls the six "domains" of the mind (negative valence, positive valence, cognitive, social, arousal/regulatory, and sensorimotor systems), each with numerous subdomains or "constructs" (the cognitive "domain," for example, is divided into constructs such as attention, memory, and language). We then seek to identify and expose, in the mentally ill person, *which faculty or faculties are failing to perform their jobs.* Nosology, here, is retooled in such a way as to actually *track the underlying pathology,* just as in Kraepelin's original vision for nosology.

There is much that can be said about the RDoC. The point that is most relevant here is that the RDoC must be seen not as a competitor to the *DSM* franchise; rather, it is the blooming of the *DSM* franchise, and the fulfillment of it.[41] It is the logical culmination of the impetus that has motivated every edition of the *DSM* from the *DSM-II* onward. It represents in pure form a return to Kant. And the animosity that proponents of the *DSM* might feel toward the RDoC should not be the animosity that one might feel toward a pesky sibling who is taking attention away from you by doing something different; rather, it is the kind of animosity that one should feel toward a pesky sibling who is taking attention away from you by doing something *better.*

15

The Darwinization of Madness

With so-called Darwinian medicine, the stage is set for a teleological reordering of psychiatry. But to understand this, we must understand how "Darwinism" itself is not an *alternative* to teleology but a naturalistic *vindication* of it (Section I). Once we see that, however, we have a prima facie sociological and historical puzzle: how is it possible that so many intelligent and earnest people have thought, and still think, that an *evolutionary* approach to psychiatry and a *biological dysfunction* approach to psychiatry will harmonize? At least theoretically, an evolutionary approach could also *undermine* a dysfunction-centered approach, because the evolutionary approach will allow us, once more, to grasp how it could be that some are truly mad by design (Section II). The real virtue of Darwinian medicine, for us, is that it helps to break the grip, the power, that *madness-as-dysfunction* exerts over contemporary psychiatric thought.

I.

Near the end of the twentieth century, a new way of conceiving of the teleology of madness emerged, thanks to the notion of a *Darwinian adaptation*. It became possible to think that depression, say, far from being a result of an inner dysfunction, could be an adaptation, could carry within it a secret purpose, drawn not from the purposiveness of an unconscious idea, or a mysterious vital force, or even the goal-directedness of the organism as such, but rather, from the workings of natural selection, from the mindless, impersonal machinery of inheritance, variation, differential fitness. This is Goldstein's vision of the organism, now atomized, fragmented, carved up into independently evolving "traits," or "mechanisms," each of which can be more or less independently modified by the power of natural selection. From an evolutionary point of view, it became possible to think that depression, the anxiety disorders, psychopathy, or even the delusions of schizophrenia could stem from mechanisms that are *performing their evolved functions perfectly*

Madness. Justin Garson, Oxford University Press. © Oxford University Press 2022.
DOI: 10.1093/oso/9780197613832.003.0016

well, that are discharging those functions quite adequately. It became possible to see in depression an archaic mechanism for navigating interpersonal conflict, or in generalized anxiety disorder a developmental mechanism for remaining hypervigilant to potential threats, or in the delusions of schizophrenia an attempt to buffer the mind from events or perceptual experiences that would otherwise totally disrupt our ability to get around in the world.

Of course, nobody maintains, today, that *all* mental disorders are adaptations. Rather, the possibility that *some* of them are violates the spirit of the Kantian approach to classification and research as it is embodied in the *DSM* and RDoC. It forces a teleological reorientation of the entire discipline. What remains astounding is the fact that so many thinkers seem to see *no contradiction between the dysfunction-centered approach to psychiatry and the evolutionary approach to psychiatry*. The failure to observe such a palpable contradiction is, in a sense, a failure that this entire book is intended to explain and, to some degree, militate against. It is the failure to see that there has always been a robust teleological "tradition"—if we dare call it that—in psychiatry, and that "Darwinian medicine" is only the latest episode in this tradition.

Darwinism and teleology: putting those terms together appears to create a tension. Didn't Darwin succeed in abolishing teleology? Wasn't the whole point of Darwinism to expunge final causes from biology? Wasn't Darwin precisely that "Newton of the blade of grass" that Kant alludes to (while denying the very possibility that the biological world would ever have its Newton)? Doesn't Darwinism show how teleology is swallowed up in mechanism? Absolutely not. In fact, the idea that "Darwinism trumps teleology" is a fundamentally *un-Darwinian* doctrine: Darwin freely spoke of the *ends*, *purposes*, and *functions* of traits. His friend Asa Grey once wrote of "Darwin's great service to natural science in bringing back to it Teleology." What is more interesting is Darwin's approving response: "What you say about Teleology pleases me especially and I do not think anyone else has ever noticed the point."[1]

But how does Darwinism bring back teleology to natural science? The thinker who, *intellectually*, did the most to advance this synthesis is all but unknown today, the French philosopher of science Edmond Goblot (1858–1935).[2] At the beginning of the twentieth century, there were two main viewpoints about the relationship between Darwinism and teleology; that clash of viewpoints was on display in a debate that took place at that time between the biologists Charles Richet and Sully Prudhomme.[3] Richet's point of view was that it is quite obvious that there is teleology in biology; biology would be fundamentally impossible if scientists were forced to stop inquiring after the "function" or "purpose" of a trait; but, he reasoned, Darwinian natural selection cannot explain the purposiveness or goal-directedness of the organism or its parts; therefore, there must be some principle, *superadded* to natural selection, that explains teleology. This, for Richet, was the mysterious vital force. Prudhomme, in contrast, held that teleology has been eliminated from biology *thanks to* Darwinian natural selection. We no longer need to talk in any literal manner about purpose, design, or function; we must speak instead of what a trait was "selected for," what natural selection shaped it for. For Prudhomme, Darwinism simply replaces teleology.

In light of this discussion, Goblot worked out, rigorously, a new conception of teleology that is *grounded* in Darwinian natural selection, in a paper fittingly entitled "La finalité sans intelligence." The reason we can speak of the purpose, end, or function of the nectar glands of the flower is precisely because *those glands were shaped by natural selection for that end*:

> But if it happens that an individual character is an *advantage*, natural selection will make of it a species character, and that because it is an advantage. Hence again there is finality, but finality without intelligence. . . . It is easy to see that these examples [e.g., "the function of nectar glands in flowers is to attract insects"] answer to the definition of finality, for the consequent is the *raison d'être* of the antecedents. Cross-fertilization exists because it causes greater fecundity; nectar glands, large or brilliant corollas, perfumes exist because they have the effect of attracting insects. . . . It would not be exact to say that the effect is here the cause of its cause, but it is true to say that it is the reason for it; the existence of the cause is explained by the effects that it produces.[4]

A seeming problem with Goblot's solution is that we have simply replaced an appeal to supernatural or vital forces with a *bad metaphor*. Natural selection is a source of purpose, one might think, because it is a source of design. Natural selection designs our traits in just the manner that a leatherworker designs a saddle. Surely, this thought represents a complete misinterpretation of the very doctrine of evolution by natural selection, which, we are repeatedly told, requires no consciousness, foresight, or intelligence.

For Goblot, the road to fusing Darwinism and teleology is not through the lame metaphor of design; it is through what the philosopher Larry Wright later came to call the "consequence etiology."[5] *Often enough*, when a biologist assigns a function to a trait, she is trying to explain *why the trait exists*: when I say that the function of the zebra stripes is to deter biting flies, I am trying to say why zebras have stripes. That raises a puzzle: how is it possible for the *effect* of a trait to explain the *existence* of that very trait? Natural selection enters as the solution to that puzzle. If, in the past, a trait was selected for deterring flies, then it is true to say that the trait exists now *because* it deters flies. Natural selection grounds teleological explanation by showing how it is possible to explain the existence of a trait via one of its beneficial effects. Selectionist explanations support, in the strictest sense of the term, teleological explanations. Selectionist explanations *are* teleological explanations. This was Goblot's insight in 1900. And though Goblot was promptly forgotten, his insight was rediscovered 73 years later by Larry Wright, and elaborated into the selected-effects theory of function, which we owe (in its modern form) to Karen Neander and Ruth Millikan.[6]

In light of the intimate relationship between teleology and natural selection, the only remaining puzzle is this: why have so many theorists of madness—philosophers and psychiatrists alike—thought that an evolutionary approach to madness and a dysfunction-centered approach to madness could be seamlessly fused? How did so many people fail to appreciate the tension between the two approaches?

Here is *one, quite partial*, answer to our question: the *DSM* franchise was so successful at abolishing every last remnant of teleology from madness that it was difficult to even entertain the idea that there could be an alternative to a dysfunction-centered account. And if there is no alternative to a dysfunction-centered account, then we surely cannot anticipate that Darwinism will undermine or abolish it. We have already seen the first major attempt to synthesize a dysfunction-centered view with an evolutionary view: that is Donald Klein's answer to Barton's prompt to define "mental illness." For Klein, there is no question that mental disorders involve inner dysfunctions; dysfunction, after all, is what distinguishes mental disorders from any other kind of socially disvalued behavior. But by what right do we judge that there is a dysfunction? Who gets to decide what is functional and what is dysfunctional? For Klein, *Darwin gets to decide that*. Something in the mind or in the brain is dysfunctional, he thought, when it fails to perform its evolutionary function, that is, when it is unable to do what natural selection designed it to do. As he put it, "Can we arrive at a standard [of disease] that is not simply an expression of personal preference, *but is given to us by the biology of the situation?*"

To illustrate the point, suppose that there is a mechanism in my brain that has the evolved function of helping me empathize with other people, since empathy is a precondition of sociality and sociality is a precondition of survival in our species. In the psychopath, this mechanism is somehow disrupted—because of a "genetic" cause, or an "environmental" cause, or some combination of the two. This mechanism, in sum, cannot perform its evolved function. Therefore, the psychopath has a genuine mental disorder; psychopathy is not just a matter of "social deviance." All relativism, all social constructionism, must kneel before the altar of objective biological fact. But that raises an obvious question: *how do we know* that our minds were not designed for asociality, for depression, for delusions and hallucinations? Again, that question *could scarcely arise* after the 1980s, because the teleological perspective had been abandoned in American psychiatry.

The thinker who has most forcefully promoted this fragile synthesis between an evolutionary and a dysfunction-centered view of madness is Wakefield. Largely as a response to the attempts of psychiatrists like Spitzer and Klein, Wakefield assembled his own definition of mental disorder in his landmark 1992 paper, "The Concept of Mental Disorder: On the Boundary Between Biological Facts and Social Values." There, as noted above, he promotes what he calls the "harmful dysfunction" analysis of disorder. In order for something to be a mental disorder—rather than, say, a bit of socially disapproved behavior—two things must be true of it: *first*, it must stem from some kind of inner dysfunction, and *second*, it must be sufficiently harmful (where "harm" is judged by prevailing social standards). In this respect, Wakefield's theory seems to be a happy synthesis of "naturalism" and "normativism." Whether or not a mental condition stems from a dysfunction is based on *objective biological facts*; whether or not it is deemed to be harmful is based on *social attitudes and values*. But what is a function? And what is a dysfunction? Wakefield, adopting the theory of function initially envisioned by Goblot, tells us that the function of a trait is whatever it was shaped for by natural selection: "Because natural selection is the only known means by which an effect can explain a naturally occurring mechanism that provides it, evolutionary explanations presumably underlie all correct ascriptions of natural functions. Consequently, an evolutionary approach to personality and mental functioning is central to an understanding of psychopathy."[7] His view can be read, therefore, as a philosophically richer version of Klein's.

The most extraordinary thing about his view is that he holds that this statement, *disorders are harmful dysfunctions*, is part of a correct conceptual analysis of "disorder" itself. In other words, it is a *tautology* to say that mental disorders are harmful dysfunctions. Anyone who says, "I think depression is a mental disorder, but there's no dysfunction; rather, it was shaped by evolution to help us resolve interpersonal conflict," or "it was designed by God to humble us," is not saying something false, he is saying something incoherent. It would be analogous to saying, "I think most bachelors are unmarried men, but I suppose some of them are married, too."

A question that I have pondered frequently, and yet failed to fully answer, is: *how is this possible?* How is it possible, sociologically and historically? What does it mean, from a social and historical point of view, that an idea that represented mainstream psychiatric thinking in 1952—namely, that all forms of madness are merely strategies for dealing with unpleasant situations—becomes seen a mere 40 years later not as dubious or false, but conceptually incoherent? The fact that the *DSM-II* and future editions did a very good job of expunging reference to teleology is certainly part of the answer to that. But that cannot be the entire answer. That is because many of the people who eventually approved Wakefield's definition, who heartily endorsed it, such as Spitzer and Klein and even Randolph Nesse, were alive and well when psychoanalysis still reigned in professional American psychiatry. They were alive and well when it was taken as a bit of conventional wisdom in mainstream American psychiatry that mental disorders were, at root, functional and not dysfunctional. It is as if, during the period from 1952 to 1992, a kind of collective amnesia descended upon mainstream American psychiatry, and all of the thinkers who advocated this teleological viewpoint, such as Freud and Goldstein and Sullivan, and even the anti-psychiatrists like Laing, were suddenly, as it were, speaking a different language. They were not saying things that were empirically false; they were saying things that were confused or nonsensical, or saying things in a language we have forgotten how to translate. I can only leave this as a question for historians and sociologists to answer.

II.

Wakefield's attempt to articulate this synthesis of an evolutionary and a dysfunction-centered account, to suggest that the dysfunction-centered perspective represents the apex of good biological theorizing, that it represents a working out of the systematic implications of modern evolutionary thought, was, intellectually speaking, dismantled almost from the moment it had been clearly formulated. 1994 witnessed the publication of Nesse and Williams' famous *Why We Get Sick: The New Science of Darwinian Medicine*. It sought to explain many different diseases not as failures of function but as adaptations. A paradigm case was morning sickness. Morning sickness, they urged, should never be treated as a disease; it is an adaptation to prevent the fetus from exposure to toxins in the mother's diet. Just as fever was once thought to be a disease and is now thought to be a component of the healing process, so too, they think, our understanding of many diseases will be transformed in this way.

Nesse and Williams devote one chapter to psychiatry and to the theory that some mental illnesses are adaptations. In a passage that Pinel would have admired, they write:

> As is the case for the rest of medicine, many psychiatric symptoms turn out not to be diseases themselves but defenses akin to fever and cough. Furthermore, many of the genes that predispose to mental disorders are likely to have fitness benefits, many of the environmental factors that cause mental disorders are likely to be novel aspects of modern life, and many of the more unfortunate aspects of human psychology are not flaws but design compromises.[8]

True, the evolutionary perspective in psychiatry is not committed to the idea that *all* mental disorders are adaptations; some of them may very well be dysfunctions of innate mental mechanisms. Nonetheless, the evolutionary perspective encourages us, as did Freud, to consider the prospect that some mental disorders that strike us, at first glance, as dysfunctions contain a hidden purpose or end. It invites us, in other words, to *reinstate* a certain measure of purposiveness and goal-directedness into madness. For example, Nesse and Williams develop an adaptationist account of depression. Depression, they reason, probably evolved as a mechanism for helping us detach from unrealistic life goals:

> "Low mood" keeps us from jumping precipitously to escape temporary difficulties, but as difficulties continue and grow and our life's energies are progressively wasted, this emotion helps to disengage us from a hopeless

enterprise so that we can consider alternatives. Therapists have long known that many depressions go away only after a person finally gives up some long-sought goal and turns his or her energies in another direction. The capacity for high and low mood seems to be a mechanism for adjusting the allocation of resources as a function of the propitiousness of current opportunities.[9]

Phobias are another obvious example of a potential adaptation:

Some novel situations, especially flying and driving, do often cause phobias. In both cases, the fear has been prepared by eons of exposure to other dangers. Fear of flying has been prepared by the dangers associated with heights, dropping suddenly, loud noises, and being trapped in a small, enclosed place. The stimuli encountered in an automobile zooming along at sixty miles an hour are novel, but they too hark back to ancestral dangers associated with rapid movement, such as the rushing attack of a predator.[10]

Discussing phobias as adaptations raises the problem of *mismatches*. When we say that a phobia like fear of snakes is an adaptation, we do not necessarily mean that it is *currently beneficial*, today, now; what we mean is that it was beneficial in the past and that is *the reason it currently exists*. Among adaptations, there are two main classes; those that have continuing usefulness, and those that are no longer useful because the environment has changed in such a way as to render them otiose.[11] The latter are *mismatches*. A mismatch is a trait that was selected for some benefit in the past, but our environment has changed to the point where it is no longer accompanied by that benefit. Though it is a mismatch, it still admits of a teleological explanation. Snake phobias are there because they helped our ancestors back in the Pleistocene, not necessarily because they help us today.[12]

Though *Why We Get Sick* gives short shrift to madness, other theorists began to fill in that gap very quickly. This led to an explosion of adaptationist hypotheses for various mental illnesses. Nesse, as noted above, favored the hypothesis that depression is, in fact, a mechanism for helping us to detach from unrealistic life goals. Depression is not a mismatch or a vestige; it is not that it was once an adaptation but because of a changed environment it no longer is so; rather, it still serves this function. Other theorists sought to demonstrate the secret purposiveness of anxiety disorders, psychopathy, delusions, and personality disorders. Brüne provides a monumental introduction to, and synthesis of, adaptationist hypotheses of mental illness.[13] Philosophers were quick to acknowledge at least the possibility of seeing mental disorders as evolutionary mismatches, and therefore *not* dysfunctions: this is the so-called *evolutionary mismatch* critique of Wakefield.[14]

It is easy to denigrate all adaptationist theories of mental illness as mere *just-so stories*, as cheap and unverifiable pop-Darwinist reasoning. But we should resist that shallow critique, for two reasons. *First*, many of these hypotheses have successfully made the leap from the stage of armchair reasoning and have begun to incorporate new lines of evidence from the realms of cognitive neuroscience, developmental biology, and comparative ethology for support.[15] Glover, for example, gives an overview of sophisticated attempts to think of anxiety disorders as adaptations to early developmental stressors.[16] She reflected on the fact that there appears to be a correlation between early (pre- and post-natal) exposure to stress and susceptibility to anxiety disorders. The more stressors there are in the environment, as indicated by maternal stress (cortisol), the more the fetal brain calibrates its hypothalamic-pituitary-adrenal (HPA) axis in such a way as to increase its *own* stress response to future stressors. The fetus effectively utilizes the level of maternal stress, as measured by cortisol, as an indication of the number or severity of stressors in the environment it will soon be forced to inhabit.

This correlation takes on new meaning when it is incorporated into an evolutionary narrative. Why, evolutionarily, would highly stressful formative environments effectively lead the developing organism to calibrate its anxiety level accordingly? What would be the *purpose* of this arrangement? In short, a more anxious individual will be *hypervigilant* to potential threats around him. As the theorist Peter Gluckman and his colleagues suggest, we must think of the developing organism as making a kind of gamble; it must sample its current environment in order to better predict what its future environment will be like; this is what they call a *predictive adaptive response*.[17] If it experiences a high level of stress in its present environment, it makes a kind of inductive inference: *my present environment has a lot of threats in it; probably, my future environment will have a lot of threats in it, too; therefore, I had better select a high-anxiety phenotype in order to be extremely vigilant to these threats.*

Others have sought to wed this broadly adaptationist vision to cognitive neuroscience. As we have seen, one theory of the delusions of schizophrenia is that the delusions represent an adaptive response to abnormal perceptual experiences. Suppose there is some low-level perceptual abnormality—a dysfunction, if you will—that generates non-veridical perceptual experiences. I hear a burst of radio transmission in my head. One way I try to account for this unusual experience is to suppose that my dentist has implanted, while I was unconscious, some kind of radio-transmission device. I progressively elaborate and systematize a set of delusions to make sense of this abnormal experience. Recently, the neuroscientists Uhlhaas and Mishara have argued that these delusions might actually serve such a function; their function is to provide the person with a coherent framework of ideas that enable him or her to continue operating at a basic level in society.[18] The philosophers Bortolotti and Sullivan-Bissett have recently used the term "epistemic innocence" to describe the apparent *reasonableness* of some delusions, given the experiences that fostered them.[19] McKay and Dennett also explore, though they ultimately reject, the hypothesis that delusions are "design features."[20] A similar idea, recall, was also part of Solomon Snyder's dopamine hypothesis of schizophrenia. His view was that excessive dopamine production leads to massive psychological deterioration but that, if the sufferer is able to construct a system of delusions to make sense of that experience, then he or she might be able to "get by" in the world: with delusions, the patient "strive[s] for an intellectual framework in which to focus all the strange feelings that are coming over him as the psychosis develops."[21] In this, Snyder echoes the earlier view of Karl Jaspers, who held that delusions are, often enough, an epiphanic resolution of a generalized feeling of strangeness.[22] Of course, one might argue that in this case, there is still a dysfunction, namely, the perceptual dysfunction that gave rise to the delusions. No doubt that may be true, but the point here is to demonstrate how easy it is to mistake a symptom that is actually functional and adaptive for a dysfunction, how easy it is to think of a function as a dysfunction—the precise tendency that Freud repeatedly warned against.

And this leads to a *second* reason we should resist the "just-so stories" complaint that we have heard so much of. The reason that these adaptationist hypotheses are so interesting for the philosopher is not necessarily because they are *true*. The philosopher, qua philosopher, has no right to adjudicate such weighty matters; this is an "empirical question." The very coherence and plausibility of the Darwinian account imply that madness-as-dysfunction can no longer be a silent default in approaching the mad. The mental reflex that compels us, when confronting the mad, to ask, "What has gone wrong inside of you? What is going wrong inside of you?," must not merely be suspended. It must be purged, even violently so. We do not urge the rejection of madness-as-dysfunction, as if it has no proper place in our thinking; indeed, we submit that some expressions of mad resistance have erred on the opposite side by rejecting, in an a priori manner, anything that resembles a "medical model" of madness. The point is to identify and expose madness-as-dysfunction as merely one style of thinking, and to force it to coexist with other styles of thinking, to force it to be a *good citizen*.

Epilogue

We have now completed our task. We have created a new space, an open field, for "rethinking madness."

"But what shall occupy that space?"

Our *tentative and highly provisional* answer: we put nothing "there." It is true that, now that we have transcended madness-as-dysfunction, and even transcended madness-as-strategy, we have space for a new thing, *madness-as-X*. We may even feel an overwhelming intellectual compulsion to put something there. But it is possible that this very habit of thought is condemnable, because of the manner in which it posits madness as something *over there*, to be observed and discussed, as if we have emptied a cage in a zoo and now ask what animal shall be placed inside of it.

We here recall Wigan's troubling thought that, scientifically speaking, the existence of madness is not the problem; given the dual structure of the mind, it is not even particularly remarkable or surprising. What is both surprising and remarkable is sanity itself. The idea that sanity is a thing that is surprising and remarkable, a thing that deserves explanation, makes it possible to entertain the thought that madness is a *default state of being*, like inertia. It is not an accident that happens from time to time and that tragically befalls an otherwise healthy person, a promising young man or woman; it is not a thing that we can discuss, contemplate, think about from a safe distance. It is the ground from which all thinking originates. Madness is the soil from which various forms, or creatures, emerge and show themselves, before receding from view or being displaced by other creatures. And there is no power that has the authority to rise up and declare, "Madness must now, at long last, halt its mad productions, and make way for this other king, sanity." The semblance of such an authority is merely one form, or creature, that emerges from this soil before it is displaced in turn. Sanity is a transient subspecies of madness.

The starting point for future philosophical inquiry is not: *what is madness? What shall we compare it to?* But rather: *what is sanity? What shall we compare it to?*

Notes

Introduction

1. APA (1980, 6); Insel et al. (2010, 749). Another RDoC proponent, Charles Sanislow (2016, 222), expresses the same point: "Rather than beginning with diagnoses based on clinical description and then trying to connect them to mechanisms, RDoC research begins with dysfunctional mechanisms and works toward clinical symptoms."
2. Shorter (2005, 3); Decker (2013, xix).
3. Haslam (1798, 12).
4. Pinel (1800, 279).
5. Fleck (1979/1935).
6. Edelstein (1937, 222).
7. E.g., Bovet and Parnas (1993, 595). Recently, Ritunnano and her colleagues (2021) have demonstrated that madness-as-strategy can be a suitable paradigm for rethinking the nature and function of delusions.
8. See, e.g., Hall (2016); Bossewitch (2016); Curtis et al. (2000); for philosophical approaches to Mad Pride, see Hoffman (2019) and Rashed (2019).

Chapter 1: Hippocrates and the Magicians

1. Nutton (2004, 71).
2. For recent archaeological work, see https://www.archaeology.org/news/8398-200 203-greece-asclepeion-epidaurus.
3. See Laskaris (2002).
4. Hippocrates, Volume II (1923, 141). All references to Hippocrates are from the Loeb Classical Library editions.
5. Ibid., 143.
6. Ibid., 145.
7. Ibid., 148–49.
8. Ibid., 161.
9. Ibid., 177.
10. See Edelstein (1937); Lloyd (1979).
11. Hippocrates, Volume II (1923, 139), my emphasis.
12. Ibid., 141.
13. Ibid., 183.
14. Ibid., 149.

15. Jouanna (1999, 203).
16. Hippocrates, Volume I (1923, 129).
17. Hippocrates, Volume IV (1931, 437).
18. Edelstein (1937, 201). Here Edelstein is citing the scholar Allbutt.
19. Jouanna (2012, 110).
20. Ibid., 111.
21. Ibid., 107.
22. Ibid., 116.
23. Edelstein (1937, 213).
24. Hippocrates, Volume I (1923, 299).
25. See Weinfeld (1973).
26. Hippocrates, Volume I (1923, 301).
27. Ibid., 299.
28. SE (7:292). All citations to Freud refer to the volume and page number of *The Standard Edition of the Complete Psychological Works of Sigmund Freud*, edited by James Strachey and published by the Hogarth Press and the Institute of Psychoanalysis in London between 1953 and 1974.

Chapter 2: The Suffocation of the Mother

1. Jorden (1603), from the dedication.
2. Ibid., 4, my emphasis.
3. Burton (2001/1621, I.133), my emphasis.
4. Doob (1974), in her aptly titled *Nebuchadnezzar's Children*, provides a catalog of the ways the theme of madness as punishment and purgation is woven throughout medieval literature.
5. Kramer and Sprenger (1971/1486, 79).
6. James I (1597, xiv).
7. Ibid.
8. Edelstein and Edelstein (1945, 113).
9. Mark 9:29 (NKJV).
10. Schoeneman (1977) clarifies the way that demon possession and madness are distinguished in medieval times; also see Bonzol (2009).
11. James I (1597, 61).
12. Burton (2001/1621, I.143).
13. Kramer and Sprenger (1971/1486, 16).
14. Deuteronomy 18:10.
15. Jorden (1603), from the dedication.
16. Ibid., 5.
17. Ibid.
18. For a recent discussion of cessationism in relation to possession, see Bhogal (2015); also see Walker (1981, 5) and MacDonald (1991, xxvii).

19. Matthew 9:6 (ESV).
20. MacDonald (1991, xxix).
21. Jorden (1603, 2).
22. Ibid., 6.
23. Ibid.
24. Ibid.
25. MacDonald (1991, xvii).
26. Jorden (1603, 19).
27. Ibid., 25.
28. Ibid., 24.

Chapter 3: Madness as Misuse and Defect

1. Hebrews 10:12 (ESV).
2. Burton (2001/1621, I.131–32).
3. Deuteronomy 28:28 (NIV).
4. Burton (2001/1621, I.132–33).
5. Sober (1988).
6. Burton (2001/1621, I.37).
7. Ibid., I.132.
8. Ibid., I.133.
9. Ibid., I.178.
10. Galatians 6:8 (KJV).
11. Burton (2001/1621, I.136).
12. Doob (1974, 31) notes correctly that throughout medieval literature, the symptoms of madness "point to moral fault," insofar as they "indicate a certain deformity of soul, a lack of the image of God, that is associated with moral depravity." But she does not see that the specific symptoms of madness are portraits of the specific sins, nor that this resemblance is an expression of God's mercy.
13. Daniel 4:30 (ESV).
14. Daniel 4:34 (NIV).
15. Burton (2001/1621, I.8).
16. Radden (2016).
17. Burton (2001/1621, I.269).
18. Ibid., I.270.
19. Ibid., I.274.
20. Ibid., I.253.
21. Ibid., I.285.
22. Ibid., I.287.
23. Ibid.
24. Furnivall and Pollard (1904, 84).

Chapter 4: An Infinitely Wise Contrivance

1. Voltaire (1904/1764, 23).
2. Cheyne (1733, iii).
3. Ibid., ii.
4. Blackmore (1725, vi).
5. See Cooper (2010); Tsou (2020), for discussion of whether culture-bound syndromes are "real" or, alternatively, whether they are natural kinds deserving of recognition in official psychiatric nomenclature.
6. Cheyne (1733, dedication).
7. Ibid., 324.
8. Cheyne (1715, 255).
9. Ibid., 256.
10. Ibid., 258.
11. Ibid., 156.
12. Ibid., iv–v.
13. Ibid., 173.
14. See Porter's introduction to the 1991 reprint of *The English Malady*, vii.
15. The historian Anita Guerrini (2000) has done a particularly thorough job demonstrating the theistic motive animating Cheyne's work; in particular, she traces his preoccupation with intemperance as the cardinal sin to the influence of French mystic Antoinette Bourignon. Contrary to Roy Porter's claim (1987, 80–81) that *The English Malady* is a thoroughly secular text—in contrast to the religiously inspired *Anatomy of Melancholy*—she rightly emphasizes the "pervasive sense of sin" (149) underlying the book.
16. Cheyne (1733, 25).
17. Ibid.
18. Craver and Darden (2013) and Glennan (2017) are convenient reference points for approaching this transition.
19. Garson (2019).
20. Craver (2013).
21. Witness Turner (1982, 260): "While Cheyne wrote on iatromathematical topics and natural religion, the majority of his publications were accounts of and advertisements for his system of dieting which he held to be the basic remedy for mental and physical illness as well as the basis of long life."
22. Mayr (1983, 328).
23. Cheyne (1733, 157).
24. Ibid., 158–59.
25. Cheyne (1740, xxxvii).
26. Cheyne (1733, 196–200).
27. Revelation 9:20; 16:9.
28. Cheyne (1733, 364).
29. Cheyne (1740, 138–39).
30. Ibid., 11–12.
31. Ibid., 12.
32. Romans 1:20 (NIV).

33. Cheyne (1740, 12–13).
34. Cheyne (1733, 34).

Introduction to Part II

1. Kant (2006, 152) [Ak VII: 253]. The latter citation refers to the volume and paragraph number of the Akademie edition of Kant's collected works, which can be found at https://korpora.zim.uni-duisburg-essen.de/kant/verzeichnisse-gesamt.html.
2. Ibid., 110–11 [Ak VII: 216], my emphasis.

Chapter 5: A Temporary Surrogate of Reason

1. See, e.g., Scull (2015).
2. Fleck (1979/1935).
3. Foucault (2006/1961, 249).
4. Strawson (1962, 194–95). I'm grateful to Ginger Hoffman for drawing my attention to this connection.
5. Haslam (1798, 12); Rush (1835, 8–9); Spurzheim (1836, 5–6); Cox (1806, 30).
6. Kant (2011, 210) [Ak II: 264].
7. Ibid.
8. Ibid., 211 [Ak II: 266].
9. Ibid., 212 [Ak II: 267].
10. Ibid., 213 [Ak II: 267].
11. Ibid. [Ak II: 268].
12. Ibid., 214 [Ak II: 268].
13. See Frierson (2009a, 2009b) for a lucid overview.
14. Foucault (2008, 31).
15. Kant (2006, 149) [Ak VII: 251].
16. Ibid., 153 [Ak VII: 254].
17. Ibid., 152 [Ak VII: 253]. Also see Frierson (2009a, 272) for discussion.
18. Ibid., 166 [Ak VII: 266–67].
19. Ibid., 150 [Ak VII: 252].
20. Ibid., 175 [Ak VII: 274–75].
21. Ibid., 110–11 [Ak VII: 216]. Frierson (2009a, 272) emphasizes the teleological character of this passage.
22. Locke (1975/1689). Locke also touches on the topic of madness in chapter 45 of his posthumous *Of the Conduct of the Understanding*, as well as miscellaneous notes in his early medical journals. These do not substantially enhance or modify the view advanced in the *Essay*.
23. Ibid., II.11.12, p. 160.
24. Ibid., II.12.13.
25. Ibid.

26. Heyes (2018).
27. Cratsley and Samuels (2013).
28. Tabb (2019).
29. Locke (1975/1689), II.33.10.
30. Ibid., II.33.12.
31. See Tabb (2019) for an overview; her own position is the latter.
32. Kant (2011, 210) [Ak II: 270].

Chapter 6: The Mountebanks of the Mind

1. Porter (2002, 112); also see Scull (2015, chapter 7) and Shorter (1997, chapter 2) for more on the growth of the asylum in the nineteenth century.
2. Haslam (1798, 18).
3. Ibid., 2.
4. Ibid., 19.
5. Ibid., 2.
6. Ibid., 104–5.
7. Haslam (1810, 14).
8. Foote (1794, 55).
9. Ibid.
10. Haslam (1810, 16).
11. Ibid., vi.
12. Ibid., 10.
13. Ibid., 18.
14. Ibid., 19–21.
15. See Porter's (1988, xvi) introduction to the reissued volume of *Illustrations of Madness*.
16. Haslam (1810, 30).
17. Cited in Porter (1988, xxxix).
18. Porter (1987, 35).
19. Kant (2011, 215–16) [Ak II: 270].
20. Haslam (1798, 78).
21. Ibid., 33–34.
22. See Sinnott-Armstrong (2008) for an overview.
23. See Carroll (1895); Hofstadter (1979). This puzzle has birthed a massive literature; see Pavese (2021) for a recent point of entry.
24. Haslam (1798, 28–29).

Chapter 7: The Miracle of Sanity

1. Wigan (1844, 15).
2. Ibid., 4.

3. Ibid., 189.
4. Ibid., 30.
5. Harrington (1987, 29).
6. Clarke (1987, 28). Clarke also notes that there were other figures during Wigan's time, most notably Hewitt Watson and Henry Holland, who advocated for a "double-brain" thesis, but that Wigan most likely did not encounter their work until he was well into the writing of his own manuscript.
7. Wigan (1844, 35).
8. Ibid., 66.
9. Ibid., 52.
10. Ibid., 65.
11. Ibid., 224–25.
12. Ibid., 230.
13. Ibid., 146.
14. Ibid., 257.
15. Ibid., 347.
16. Ibid., 407.
17. Ibid., 313.
18. Ibid., 123; my emphasis.
19. Ibid., 128.
20. Ibid., 129.

Chapter 8: Delusion as Castle and Refuge

1. Schreber (1955/1903, 48).
2. Burton (2001/1621, 11).
3. Heinroth (1975/1818, 37; sect. 19). All references are from the Schmorak translation. For biographical information, see Marx (1990; 1991); Steinberg and Himmerich (2012).
4. Heinroth (1975/1818, 8; sect. 19).
5. Ibid., 6; sect. 13.
6. Ibid., 12; sect. 27.
7. Ibid., 16; sect. 41.
8. Ibid., 146; sect. 194.
9. Ibid., 162; sect. 204.
10. Ibid., 163; sect. 204.
11. Incidentally, Schopenhauer, too, briefly develops this theme of madness as a strategic flight from a painful memory: "If such a sorrow, such painful knowledge or reflection, is so harrowing that it becomes positively unbearable, and the individual would succumb to it, then nature, alarmed in this way, seizes on *madness* as the last means of saving life . . . thus [the mind] seeks refuge in madness from the mental suffering that exceeds its strength" (1969/1819, 192). See Brook and Young (2019) for discussion. I thank Richard Gipps for drawing this passage to my attention.
12. Bateson (1974, xiv).

13. Fromm-Reichmann (1948, 265). This paper, "Notes on the Development of Treatment of Schizophrenics by Psychoanalytic Psychotherapy," is the one that I will use here as the paragon of her work; but other papers that develop the same ideas particularly adroitly include her 1939 paper "Transference Problems in Schizophrenics," and her 1946 paper "Remarks on the Philosophy of Mental Disorder."

14. Hornstein (2000, 133).

15. E.g., see Fromm-Reichmann (1940, 132) on the cultural differences between Western Europe and the United States.

16. Fromm-Reichmann (1946, 298).

17. Fromm-Reichmann (1939, 414).

18. Ibid., 118.

19. Fromm-Reichmann (1948, 265).

20. Ibid., 273.

21. Ibid., 266.

22. Ibid., 271.

23. SE (14:74).

24. Sullivan (1962, 8).

25. Ibid., 33.

26. Meyer (1917, 444).

27. Sullivan (1956, 212).

28. Sullivan (1940, 179).

29. Sullivan (1956, 25).

30. Sullivan (1953, 161); see Chatelaine (1992) for a clear introduction to this element of Sullivan's thought.

31. Sullivan (1956, 171).

32. Ibid., 66.

33. Ibid., 24.

34. Ibid., 26.

35. Sullivan (1940, 143).

36. Sullivan (1956, 308).

37. Ibid.

38. Ibid., 351.

39. Ibid., 341.

40. Sullivan (1940, 155–56).

41. Ibid., 157.

42. Sullivan (1956, 340); also see 350 for further applications of this principle.

43. Ibid., 311–12.

Chapter 9: A Salutary Effort of Nature

1. Pinel (1800, 39). All translations are the author's own. Though the *Traité* was translated into English in 1806, the English translation is an extremely loose one and cannot be used for scholarly purposes (see Weiner 2000 for discussion).

2. Pinel (1800, 40).

3. Ibid.

4. Ibid., 262.

5. Huneman (2014) adds to this list that Pinel should be seen as one of the major innovators of the *case study* and its central role in the mental hospital.

6. Pinel (1800, 4–5).

7. Ibid., 13.

8. Ibid.

9. Ibid., 21.

10. Ibid., 24–25.

11. Ibid., 27.

12. Ibid., 99–100.

13. Foucault (2006/1961, 489).

14. Pinel (1800, 58–59).

15. Haslam (1798, 137).

16. Ibid., 126–27.

17. Pinel (1800, 37–38).

18. See Neuburger (1944); Geyer-Kordesch (1981).

19. Bynum (2001, 21).

20. Pinel (1800, 38).

21. Ibid., 38–39.

22. Ibid.

23. Ibid., 170.

24. Ibid., 169.

25. Edelstein (1937, 222).

26. Pinel (1800, 10).

27. Esquirol (1838, 113).

28. See Scull (2015) for a standard overview.

29. Pinel (1800, 104–5).

30. Ibid., 106.

Chapter 10: The Biologization of Kant

1. Kant (2011, 210) [Ak II: 270].

2. Griesinger (1867, 5–6). All passages are from the English translation of the second (1861) edition of his *Die Pathologie und Therapie der psychischen Krankheiten*.

3. See Engstrom (2003).

4. Griesinger (1867, 23).

5. Ibid.

6. Ibid., 24.

7. Ibid., 23.

8. Ibid., 60.

9. Ibid., 131–32.

10. Neander (1983).
11. Griesinger (1867,60–61).
12. Ibid., 26.
13. Ibid., 23.
14. Ibid., 42.
15. Ibid., 43.
16. Ibid., 44
17. Ibid.
18. Ibid., 48.
19. Dennett (1991, 259).
20. Griesinger (1867, 49).
21. Ibid., 49–50.
22. Ibid., 108–9; my emphasis.
23. SE (4:91).
24. Griesinger (1867, 109).
25. Kraepelin (1987, 25).
26. Piccinini and Craver (2011).
27. Kraepelin (1987, 11).
28. Ibid., 60–61.
29. Kraepelin (1912, 152–53).
30. Taylor (2007).
31. For historical overviews, see Leese (2002); Kent (2009).
32. See the anthology Ferenczi et al. (1921), with contributions from Sándor Ferenczi, Karl Abraham, Ernst Simmel, and Ernest Jones, with an introduction by Freud.
33. Kraepelin (1987, 164).
34. Ibid., 165.
35. Ibid.
36. Ibid., 166.

Introduction to Part III

1. APA (1980, 6); Insel et al. (2010, 749).

Chapter 11: The Strategies of Wish Fulfillment

1. SE (4:41–42).
2. SE (5:595)
3. SE (12:70–71).
4. SE (6:277–78).
5. SE (9:210).

6. Neander (1995) remains the canonical exposition of this "hierarchical" arrangement of functions.
7. SE (1:296).
8. SE (1:312).
9. SE (1:297).
10. SE (1:298); see also Pribram (1962).
11. SE (4:91).
12. SE (6:278).
13. For example, SE (5:595).
14. SE (13:28).
15. SE (6:1).
16. Mitchell and Black (1995, 4–5).
17. SE (2:203–4).
18. SE (2:166).
19. SE (2:214).
20. Ibid.
21. SE (3:47).
22. SE (3:48).
23. SE (3:49).
24. SE (3:52).
25. SE (3:60).
26. SE (3:184).
27. SE (3:54).
28. APA (1952, 12).
29. Ibid., 12–13.
30. Ibid., 13. Anna Freud and Wilhelm Reich (before his expulsion from the International Psychoanalytic Society) developed psychoanalytic theory to explore how character traits could represent mechanisms of defense against inner and outer threats. See Freud (1946); Reich (1972/1933).
31. Meyer (1908, 255).

Chapter 12: Madness as Creativity and Conquest

1. Goldstein (1939/1934, 43).
2. Ibid., 340.
3. This is the astute way his friend Gardner Murphy summarizes Goldstein's difference from Kurt Lewin. See Murphy (1968, 32).
4. Goldstein (1939/1934, 34–35).
5. Ibid., 340–41.
6. Ibid.
7. Ibid., 197.
8. Oyama (2000, 31).

9. Sterelny and Kitcher (1988).
10. Maslow (1943).
11. Goldstein (1939/1934, 305).
12. Ibid.
13. Maslow (1943, 373).
14. Goldstein (1939/1934, 303).
15. Goldstein (1940).
16. Canguilhem (1989/1966, 86).
17. Ibid., 100.
18. Amundson (2000, 39–40).
19. Goldstein (1940, 91).
20. Ferenczi et al. (1921, 4).

Chapter 13: From Retreat to Resistance

1. Bateson et al. (1956, 256).
2. Laing (1967, 93–94).
3. Ibid., 101.
4. Ibid., 85.
5. Ibid., 49.
6. Ibid., 107.
7. Ibid., 98.
8. Deleuze and Guattari (2009/1972, 131).
9. Ibid., 76–77.
10. Ibid., 19–20; my emphasis.
11. Ibid., 245.
12. Cooper (1967, 2).
13. Laing (1967, 104).
14. Harris (2012).
15. Laing (1967, 106).
16. Lilly (1968).
17. Higgs (2013).
18. Garson (2017).
19. "Medicine: Artificial Psychoses," *Time*, December 19, 1955, 62; also see Dyck 2008.
20. Woolley and Shaw (1954, 587–88); also see Gaddum (1954).
21. Rasmussen (2008).
22. See Garson (2017).
23. Ibid.
24. Ibid., and references therein.
25. Smith (1969, 188).
26. Toohey (2020, 142).
27. Angrist and Gershon (1969, 205).

28. Snyder et al. (1974, 1252).
29. Angrist and Gershon (1970, 102).
30. Ibid., 106.
31. Snyder (1973, 66).

Chapter 14: Confronting the Wounded Animal

1. See Decker (2013) on the composition of the *DSM-III* Task Force.
2. APA (1968, vii); emphasis mine.
3. Ibid., viii.
4. Ibid., ix.
5. Garmezy (1978, 6).
6. APA (1968, viii–ix).
7. Meyer (1908, 255).
8. Millon (1986, 35).
9. APA (1968, 48).
10. Ibid., 39.
11. Ibid., 33
12. APA (1980, 6). Cooper (2020) gives a thoughtful discussion of how the definition of mental disorder has changed from the *DSM-III* to the *DSM-5*.
13. APA (1987, xxii).
14. Ibid.
15. APA (1980, 189).
16. Ibid., 42, 55.
17. Kendell (1986, 41).
18. Klein (1978, 49).
19. Kirk and Kutchens (1992, 80).
20. See, e.g., Kirk and Kutchens (1992); Millon (1986); Bolton (2008).
21. Spitzer, Sheehy, and Endicott (1977, 3–4).
22. Ibid., 4; my emphasis.
23. Ibid., 5–6.
24. Kendell (1975, 306).
25. Ibid., 307.
26. Ibid., 312.
27. Ibid.
28. Ibid., 311.
29. Ibid., 313.
30. Ibid., 313–14.
31. Spitzer and Endicott (1978, 19).
32. Ibid., 25.
33. Wakefield (1993) reaches a similar conclusion in his critique of Spitzer's definition.
34. Millon (1986, 44).

35. The concept of the "sick role" is developed in Parsons (1951).

36. Klein (1978, 50).

37. Ibid., 51.

38. Wakefield (1992, 383).

39. See Tabb (2015); Tsou (2016); Faucher and Goyer (2016); Bluhm (2017); Hoffman and Zachar (2017), for careful analyses of the relationship between RDoC and the *DSM* franchise.

40. Insel et al. (2010, 749).

41. This interpretation is consistent with the point of view of David Kupfer, chair of the *DSM-5* Task Force, who notes that RDoC represents a "logical extension" of the *DSM*'s goal of grouping disorders by "underlying pathophysiological similarities" (Kupfer and Regier 2011). I thank Luc Faucher for drawing this paper to my attention.

Chapter 15: The Darwinization of Madness

1. Cited in Lennox and Kampourakis (2013, 437).

2. Garson (forthcoming a).

3. Prudhomme and Richet (1902).

4. Goblot (1900, 402–3); all translations from the French are my own.

5. Wright (1976, 116).

6. Neander (1983); Millikan (1984).

7. Wakefield (1992, 383).

8. Nesse and Williams (1994, 209).

9. Ibid., 217; also see Nesse (2000).

10. Ibid., 215.

11. Murphy (2005); Faucher and Blanchette (2011).

12. Snake phobias are almost certainly an adaptation; see Isbell (2009).

13. Brüne (2008).

14. See, e.g., Lilienfeld and Marino (1995, 416); Richters and Hinshaw (1999, 442); Woolfolk (1999, 662); Murphy and Stich (2000, 81–84).

15. Garson (forthcoming b).

16. Glover (2011).

17. Gluckman et al. (2005); see Garson (2015, Chapter 8) and Garson (2021) for discussion.

18. Uhlhaas and Mishara (2007).

19. Bortolotti (2015); Sullivan-Bissett (2018); also see Radden (2010).

20. McKay and Dennett (2009).

21. Snyder (1973, 66).

22. Jaspers (1997/1913, 98).

References

American Psychiatric Association. 2013. *Diagnostic and Statistical Manual of Mental Disorders: DSM-5*. Washington, DC: American Psychiatric Association.

American Psychiatric Association. 1987. *Diagnostic and Statistical Manual of Mental Disorders: DSM-III-R*. Washington, DC: American Psychiatric Association.

American Psychiatric Association. 1980. *Diagnostic and Statistical Manual of Mental Disorders: DSM-III*. Washington, DC: American Psychiatric Association.

American Psychiatric Association. 1968. *Diagnostic and Statistical Manual of Mental Disorders: DSM-II*. Washington, DC: American Psychiatric Association.

American Psychiatric Association. 1952. *Diagnostic and Statistical Manual of Mental Disorders*. Washington, DC: American Psychiatric Association.

Amundson, R. 2000. Against normal function. *Studies in History and Philosophy of Biological and Biomedical Sciences* 31: 33–53.

Angrist, B. M., and Gershon, S. 1970. The phenomenology of experimentally induced amphetamine psychosis—preliminary observations. *Biological Psychiatry* 2: 95–107.

Angrist, B. M., and Gershon, S. 1969. Amphetamine abuse in New York City—1966 to 1968. *Seminars in Psychiatry* 1 (2): 195–207.

Bateson, G. (ed.). 1974. *Perceval's Narrative: A Patient's Account of his Psychosis, 1830–1832*. New York: Morrow.

Bateson, G., Jackson, D. D., Haley, J., and Weakland, J. 1956. Toward a theory of schizophrenia. *Behavioral Science* 1: 251–64.

Bhogal, H. 2015. Miracles, cessationism, and demonic possession: The Darrell controversy and the parameters of preternature in early modern English demonology. *Preternature: Critical and Historical Studies on the Preternatural* 4: 152–80.

Blackmore, R. 1725. *A Treatise of the Spleen and Vapours*. London: J. Pemberton.

Bluhm, R. 2017. The need for new ontologies in psychiatry. *Philosophical Explorations* 20: 146–59.

Bolton, D. 2008. *What Is Mental Disorder?* Oxford: Oxford University Press.

Bonzol, J. 2009. The medical diagnosis of demonic possession in an early modern English community. *Parergon* 26 (1): 115–40.

Boorse, C. 2002. A rebuttal on functions. In *Functions: New Essays in the Philosophy of Psychology and Biology*, ed. A. Ariew, R. Cummins, and M. Perlman, 63–112. Oxford: Oxford University Press.

Bortolotti, L. 2015. The epistemic innocence of motivated delusions. *Consciousness and Cognition* 33: 490–99.

Bossewitch, J. S. 2016. Dangerous gifts: Towards a new wave of mad resistance. Dissertation, Columbia University.

Bovet, P., and Parnas, J. 1993. Schizophrenic delusions: A phenomenological approach. *Schizophrenia Bulletin* 19: 579–97.

Brook, A., and Young, C. 2019. Schopenhauer and Freud. In *The Oxford Handbook of Philosophy and Psychoanalysis*, ed. R. G. T. Gipps and M. Lacewing, 63–82. Oxford: Oxford University Press.

Brüne, M. 2008. *Textbook of Evolutionary Psychiatry: The Origins of Psychopathy.* Oxford: Oxford University Press.

Burton, R. 2001/1621. *The Anatomy of Melancholy.* New York: New York Review of Books.

Bynum, W. F. 2001. Nature's helping hand. *Nature* 414: 21.

Canguilhem, G. 1989/1966. *The Normal and the Pathological.* New York: Zone Books.

Carroll, L. 1895. What the tortoise said to Achilles. *Mind* 4: 278–80.

Chatelaine, L. L. 1992. *Good Me, Bad Me, Not Me: Harry Stack Sullivan: An Introduction to His Thought.* Dubuque, IA: Kendall.

Cheyne, G. 1740. *An Essay on Regimen.* London: C. Rivington.

Cheyne, G. 1733. *The English Malady.* London: G. Strahan.

Cheyne, G. 1715. *Philosophical Principles of Religion: Natural and Revealed.* London: G. Strahan.

Clarke, B. F. L. 1987. *Arthur Wigan and the Duality of the Mind.* Cambridge: Cambridge University Press.

Cooper, D. 1967. *Psychiatry and Anti-Psychiatry.* London: Tavistock.

Cooper, R. 2020. The concept of disorder revisited: Robustly value-laden despite change. *Aristotelian Society Supplementary Volume* 94: 141–61.

Cooper, R. 2010. Are culture-bound syndromes as real as universally-occurring disorders? *Studies in History and Philosophy of Biological and Biomedical Sciences* 41: 325–32.

Cox, J. M. 1806. *Practical Observations on Insanity.* 2nd edition. London: J. Murray.

Cratsley, K., and Samuels, R. 2013. Cognitive science and explanations of psychopathology. In *The Oxford Handbook of Philosophy and Psychiatry,* ed. K. W. M. Fulford, 413–33. Oxford: Oxford University Press.

Craver, C. 2013. Functions and mechanisms: A perspectivalist view. In *Function: Selection and Mechanisms,* ed. P. Huneman, 133–58. Dordrecht: Springer.

Craver, C. F., and Darden, L. 2013. *In Search of Mechanisms: Discoveries Across the Life Sciences.* Chicago: University of Chicago Press.

Curtis, T., et al. (eds.). *Mad Pride: A Celebration of Mad Culture.* N.p.: Chipmunkapublishing.

Decker, H. S. 2013. *The Making of DSM-III.* Oxford: Oxford University Press.

Deleuze, G., and Guattari, F. 2009/1972. *Anti-Oedipus: Capitalism and Schizophrenia.* New York: Penguin.

Dennett, D. 1991. *Consciousness Explained.* New York: Little, Brown.

Doob, P. B. 1974. *Nebuchadnezzar's Children: Conventions of Madness in Middle English Literature.* New Haven, CT: Yale University Press.

Dyck, E. 2008. *Psychedelic Psychiatry: LSD from Clinic to Campus.* Baltimore: Johns Hopkins University Press.

Edelstein, E. J., and Edelstein, L. 1945. *Asclepius: A Collection and Interpretation of the Testimonies.* Baltimore: Johns Hopkins University Press.

Edelstein, L. 1937. Greek medicine in its relation to religion and magic. *Bulletin of the Institute of the History of Science* 5: 201–46.

Engstrom, E. J. 2003. *Clinical Psychiatry in Imperial Germany: A History of Psychiatric Practice.* Ithaca, NY: Cornell University Press.

Esquirol, E. 1838. *Des maladies mentales.* Brussels: Meline, Cans.

Faucher, L., and Blanchette, I. 2011. Fearing new dangers: Phobias and the cognitive complexity of human emotions. In *Maladapting Minds: Philosophy, Psychiatry, and Evolutionary Theory,* ed. R. Adriaens and A. De Block, 33–64. Oxford: Oxford University Press.

Faucher, L., and Goyer, S. 2016. RDoC: Thinking outside the DSM box without falling into a reductionist trap. In *The DSM-5 in Perspective*, ed. S. Demazeux and P. Singy, 199–224. Dordrecht: Springer.

Ferenczi, S., et al. 1921. *Psycho-Analysis and the War Neuroses*. London: International Psycho-Analytical Press.

Fleck, L. 1979/1935. *Genesis and Development of a Scientific Fact*. Chicago: University of Chicago.

Foote, S. 1794. *The Devil upon Two Sticks*. London: Lowndes and Bladon.

Foucault, M. 2008. *Introduction to Kant's Anthropology*. Los Angeles: Semiotext(e).

Foucault, M. 2006/1961. *History of Madness*. New York: Routledge.

Freud, A. 1946. *The Ego and the Mechanisms of Defense*. New York: International Universities Press.

Freud, S., Strachey, J., Freud, A., and Rothgeb, C. L. 1953. *The Standard Edition of the Complete Psychological Works of Sigmund Freud*. London: Hogarth Press and the Institute of Psycho-Analysis.

Frierson, P. 2009a. Kant on mental disorder. Part 1: An overview. *History of Psychiatry* 20: 269–89.

Frierson, P. 2009b. Kant on mental disorder. Part 2: Philosophical implications of Kant's account. *History of Psychiatry* 20: 290–310.

Fromm-Reichmann, F. 1948. Notes on the development of treatment of schizophrenics by psychoanalytic psychotherapy. *Psychiatry* 11: 263–73.

Fromm-Reichmann, F. 1946. Remarks on the philosophy of mental disorder. *Psychiatry* 9: 293–308.

Fromm-Reichmann, F. 1940. Notes on the mother role in the family group. *Bulletin of the Menninger Clinic* 4: 132–48.

Fromm-Reichmann, F. 1939. Transference problems in schizophrenics. *Psychoanalytic Quarterly* 8: 412–26.

Furnivall, F. J., and Pollard, A. W. 1904. *The Macro Plays: Mankind; Wisdom; The Castle of Perseverance*. London: Kegan Paul, Trench, Trübner.

Gaddum, J. H. 1954. Drugs antagonistic to 5-hydroxytryptamine. In *Ciba Foundation Symposium on Hypertension: Humoral and Neurogenic Factors*, ed. G. E. W. Wolstenholme and M. P. Cameron, 75–77, 85–90. Boston: Little, Brown.

Garmezy, N. 1978. Never mind the psychologists: Is it good for the children? *The Clinical Psychologist* 31: 1–6.

Garson, J. 2015. *The Biological Mind: A Philosophical Introduction*. London: Routledge.

Garson, J. 2017. A "model schizophrenia": Amphetamine psychosis and the transformation of American psychiatry. In *Technique in the History of the Brain and Mind Sciences*, ed. S. Casper and D. Gavrus, 202–28. Rochester: University of Rochester Press.

Garson, J. 2019. *What Biological Functions Are and Why They Matter*. Cambridge: Cambridge University Press.

Garson, J. 2021. The developmental plasticity challenge to Wakefield's view, in *Defining Mental Disorder: Jerome Wakefield and His Critics*, eds. L. Faucher and D, Forest, 335–352. Cambridge, MA: MIT Press.

Garson, J. Forthcoming a. Edmond Goblot's (1858–1935) selected effects theory of function: A reappraisal. *Philosophy of Science*.

Garson, J. Forthcoming b. *The Biological Mind: A Philosophical Introduction*, 2nd edition. London: Routledge.

Geyer-Kordesch, J. 1981. Fevers and other fundamentals: Dutch and German medical explanations c. 1680 to 1730. *Medical History* 1 (suppl): 99–120.

Glennan, S. 2017. *The New Mechanical Philosophy*. Oxford: Oxford University Press.

Glover, V. 2011. Prenatal stress and the origins of psychopathology: An evolutionary perspective. *Journal of Child Psychology and Psychiatry* 54 (2): 356–67.

Gluckman, P., et al. 2005. Predictive adaptive responses and human evolution. *Trends in Ecology and Evolution* 20: 527–33.

Goblot, E. 1900. La finalité sans intelligence. *Revue de Métaphysique et de Morale* 8: 393–406.

Goldstein, K. 1940. *Human Nature in the Light of Psychopathology*. Cambridge, MA: Harvard University Press.

Goldstein, K. 1939/1934. *The Organism: A Holistic Approach to Biology Derived from Pathological Data in Man*. New York: American Book Company.

Griesinger, W. 1867. *Mental Pathology and Therapeutics*, trans. J. L. Robertson and J. Rutherford. London: The New Sydenham Society.

Grinker, R. R. 2021. *Nobody's Normal: How Culture Created the Stigma of Mental Illness*. New York: W. W. Norton.

Guerrini, A. 2000. *Obesity and Depression in the Enlightenment: The Life and Times of George Cheyne*. Norman: University of Oklahoma Press.

Hall, W. 2016. *Outside Mental Health: Voices and Visions of Madness*. N.p.: Madness Radio.

Harrington, A. 1987. *Medicine, Mind, and the Double Brain*. Princeton, NJ: Princeton University Press.

Harris, D. 2012. *The Residents: The Experimental Community of R. D. Laing, Kangsley Hall, 1965–1970*. N.p.: D. Harris.

Haslam, J. 1810. *Illustrations of Madness*. London: G. Hayden.

Haslam, J. 1798. *Observations on Insanity*. London: J. Hatchard.

Heinroth, J. C. A. 1975/1818. *Textbook of Disturbances of Mental Life*, trans. J. Schmorak. Baltimore: Johns Hopkins University Press.

Heyes, C. 2018. *Cognitive Gadgets: The Cultural Evolution of Thinking*. Cambridge, MA: Harvard University Press.

Higgs, J. 2013. *I Have America Surrounded: The Life of Timothy Leary*. London: Thistle.

Hippocrates. 1931. *Volume IV: Nature of Man*. Translated by W. H. S. Jones. Cambridge, MA: Harvard University Press.

Hippocrates. 1923. *Volume I: Ancient Medicine*. Translated by W. H. S. Jones. Cambridge, MA: Harvard University Press.

Hippocrates. 1923. *Volume II: Prognostic*. Translated by W. H. S. Jones. Cambridge, MA: Harvard University Press.

Hofstadter, D. 1979. *Gödel, Escher, Bach: An Eternal Golden Brain*. New York: Basic.

Hoffman, G. A. 2019. Public mental health without the health? Challenges and contributions from the Mad Pride and neurodiversity paradigms. *Developments in Neuroethics and Bioethics* 2: 289–326.

Hoffman, G. A., and Zachar, P. 2017. RDoC's metaphysical assumptions: Problems and promises. In *Extraordinary Science and Psychiatry*, ed. J. Poland and S. Tekin, 59–86. Cambridge, MA: MIT Press.

Hornstein, G. A. 2000. *To Redeem One Person Is to Redeem the World: The Life of Frieda Fromm-Reichmann*. New York: The Free Press.

Huneman, P. 2014. Writing the case—Pinel as psychiatrist. *Republics of Letters: A Journal for the Study of Knowledge, Politics, and the Arts* 3 (2): 1–28.

Insel, T. R., et al. 2010. Research Domain Criteria (RDoC): Toward a new classification framework for research on mental disorders. *American Journal of Psychiatry* 167: 748–51.

Isbell, L. A. 2009. *The Fruit, the Tree, and the Serpent: Why We See So Well.* Cambridge, MA: Harvard University Press.

James I, King of England. 1597. *Daemonologie in Forme of a Dialogue, Diuided into Three Bookes.* Edinburgh: Robert Walde-graue.

Jaspers, K. 1997/1913. *General Psychopathology, Volume 1.* Translated by J. Hoenig and M. W. Hamilton. Baltimore: Johns Hopkins University Press.

Jorden, E. 1603. *A Briefe Discourse of a Disease Called the Suffocation of the Mother.* London: John Windet.

Jouanna, J. 2012. *Greek Medicine from Hippocrates to Galen.* Translated by Neil Allies. Leiden: Brill.

Jouanna, J. 1999. *Hippocrates.* Translated by M. B. DeBevoise. Baltimore: Johns Hopkins University Press.

Kant, I. 2011. Essay on the maladies of the head. Translated by Holly Wilson. In *Observations on the Feeling of the Beautiful and Sublime and Other Writings*, ed. P. Frierson and P. Guyer, 205–17. Cambridge: Cambridge University Press.

Kant, I. 2006. *Anthropology from a Pragmatic Point of View.* Translated by Robert B Louden. Cambridge: Cambridge University Press.

Kendell, R. E. 1975. The concept of disease and its implications for psychiatry. *British Journal of Psychiatry* 127: 305–15.

Kendell, R. E. 1986. "What are mental disorders?" In *Issues in Psychiatric Classification: Science, Practice, and Social Policy*, ed. A. M. Freedman, R. Brotman, I. Silverman, and D. Huston, 23–45. New York: Human Sciences Press.

Kent, S. K. 2009. *Aftershocks: Politics and Trauma in Britain: 1918–1931.* New York: Palgrave Macmillan.

Kirk, S. A., and Kutchins, H. 1992. *The Selling of DSM.* New York: Aldine de Gruyter.

Klein, D. 1978. A proposed definition of mental illness. In *Critical Issues in Psychiatric Diagnosis*, ed. R. L. Spitzer and D. L. Klein, 41–71. New York: Raven Press.

Kraepelin, E. 1987. *Memoirs.* Translated by Cheryl Wooding-Deane. Berlin: Springer.

Kraepelin, E. 1912. *Clinical Psychiatry: A Textbook for Students and Physicians.* Translated from the 7th edition by A. Ross Diefendorf. London: Macmillan.

Kramer, H., and Sprenger, J. 1971/1486. *The Malleus Maleficarum.* New York: Dover.

Kupfer, D. J., and Regier, D. A. 2011. Neuroscience, clinical evidence, and the future of psychiatric classification in *DSM-5. American Journal of Psychiatry* 168: 672–74.

Laing, R. D. 1967. *The Politics of Experience.* New York: Random House.

Laskaris, J. 2002. *The Art Is Long: On the Sacred Disease and the Scientific Tradition.* Leiden: Brill.

Leese, P. 2002. *Shell Shock: Traumatic Neurosis and the British Soldiers of the First World War.* New York: Palgrave Macmillan.

Lennox, J. G., and Kampourakis, K. 2013. Biological teleology: The need for history. In *The Philosophy of Biology: A Companion for Educators*, ed. K. Kampourakis, 421–54. Dordrecht: Springer.

Lilienfeld, S. O., and L. Marino. 1995. Mental disorder as a Roschian concept: A critique of Wakefield's "harmful dysfunction" analysis. *Journal of Abnormal Psychology* 104: 411–20.

Lilly, J. C. 1968. *Programming and Metaprogramming in the Human Biocomputer*. New York: Julian Press.

Lloyd, G. E. R. 1979. *Magic, Reason and Experience: Studies in the Origin and Development of Greek Science*. Cambridge: Cambridge University Press.

Locke, J. 1975/1689. *An Essay Concerning Human Understanding*. Oxford: Clarendon Press.

Macdonald, M. (ed.) 1991. *Witchcraft and Hysteria in Elizabethan London*. London: Routledge.

Marx, O. M. 1991. German romantic psychiatry: Part 2. *History of Psychiatry* 2: 1–25.

Marx, O. M. 1990. German romantic psychiatry: Part 1. *History of Psychiatry* 1: 351–81.

Maslow, A. 1943. A theory of human motivation. *Psychological Review* 50: 370–96.

Mayr, E. 1983. How to carry out the adaptationist program? *American Naturalist* 121: 324–34.

McKay, R. T., and Dennett, D. 2009. The evolution of misbelief. *Behavioral and Brain Sciences* 32: 493–510.

Meyer, A. 1917. The approach to the investigation of dementia praecox. *The Chicago Medical Recorder* 39: 441–45.

Meyer, A. 1908. The problems of mental reaction-types, mental causes and diseases. *The Psychological Bulletin* 5: 245–61.

Millikan, R. G. 1984. *Language, Thought, and Other Biological Categories*. Cambridge, MA: MIT Press.

Millon, T. 1986. On the past and future of the *DSM-III*: Personal recollections and projections. In *Contemporary Directions in Psychopathology: Toward the DSM-IV*, ed. T. Millon and G. L. Klerman, 29–70. New York: Guilford Press.

Mitchell, S. A., and Black, M. J. 1995. *Freud and Beyond: A History of Modern Psychoanalytic Thought*. New York: Basic Books.

Murphy, D. 2005. Can evolution explain insanity? *Biology and Philosophy* 20: 745–66.

Murphy, D., and Stich, S. 2000. Darwin in the madhouse: Evolutionary psychology and the classification of mental disorders. In *Evolution and the Human Mind: Modularity, Language, and Meta-Cognition*, ed. P. Carruthers and A. Chamberlain, 62–92. Cambridge: Cambridge University Press.

Murphy, G. 1968. Personal impressions of Kurt Goldstein. In *The Reach of Mind: Essays in Memory of Kurt Goldstein*, ed. M. Simmel, 31–34. Berlin: Springer-Verlag.

Neander, K. 1995. Misrepresenting and malfunctioning. *Philosophical Studies* 79: 109–41.

Neander, K. 1983. Abnormal psychobiology. Dissertation, La Trobe University.

Nesse, R. M. 2000. Is depression an adaptation? *Archives of General Psychiatry* 57: 14–20.

Nesse, R. M., and Williams, G. C. 1994. *Why We Get Sick: The New Science of Darwinian Medicine*. New York: Times Books.

Neuburger, M. 1944. An historical survey of the concept of nature from a medical viewpoint. *Isis* 35: 16–28.

Nutton, V. 2004. *Ancient Medicine*. London: Routledge.

Oyama, S. 2000. *The Ontogeny of Information: Developmental Systems and Evolution*. 2nd edition. Durham, NC: Duke University Press.

Parsons, T. 1951. *The Social System*. Glencoe, IL: Free Press.

Pavese, C. 2021. Lewis Carroll's regress and the presuppositional structure of arguments. *Linguistics and Philosophy*. [Epub ahead of print] DOI: https://doi.org/10.1007/s10988-020-09320-9.

Piccinini, G., and Craver, C. 2011. Integrating psychology and neuroscience: Functional analyses as mechanism sketches. *Synthese* 183: 283–311.

Pinel, P. 1800. *Traité médico-philosophique sur l'aliénation mentale, ou la manie.* Paris: Richard, Caille, et Ravier Libraires.

Porter, R. 2002. *Madness: A Brief History.* Oxford: Oxford University Press.

Porter, R. 1991. Introduction. In *George Cheyne: The English Malady (1733).* London: Routledge.

Porter, R. 1988. Introduction. In *Illustrations of Madness.* London: Routledge.

Porter, R. 1987. *Mind Forg'd Manacles: A History of Madness in England from the Restoration to the Regency.* Cambridge, MA: Harvard University Press.

Pribram, K. H. 1962. The neuropsychology of Sigmund Freud. In *Experimental Foundations of Clinical Psychology*, ed. A. Bachrach, 442–68. New York: Basic Books.

Prudhomme, S., and Richet, C. 1902. *Le problème des causes finales.* Paris: Félix Alcan.

Radden, J. 2016. *Melancholic Habits: Burton's Anatomy and the Mind Sciences.* Oxford: Oxford University Press.

Radden, J. 2010. *On Delusion.* London: Routledge.

Rashed, M. A. 2019. *Madness and the Demand for Recognition.* Oxford: Oxford University Press.

Rasmussen, N. 2008. *On Speed: The Many Lives of Amphetamines.* New York: New York University Press.

Reich, W. 1972/1933. *Character Analysis.* New York: Farrar, Straus, and Giroux.

Richters, J., and Hinshaw, S. 1999. The abduction of disorder in psychiatry. *Journal of Abnormal Psychiatry* 108: 438–45.

Ritunnano, R., et al. 2021. Finding order within the disorder: A case study exploring the meaningfulness of delusions. *BJPsych Bulletin* [Epub ahead of print] DOI: https://doi.org/10.1192/bjb.2020.151.

Rush, B. 1835. *Medical Inquiries and Observations upon the Diseases of the Mind.* 5th edition. Philadelphia: Grigg and Elliot.

Sanislow, C. A. 2016. Updating the research domain criteria. *World Psychiatry* 15 (3): 222–23.

Schoeneman, T. J. 1977. The role of mental illness in the European witch hunts of the sixteenth and seventeenth centuries: An assessment. *Journal of the History of the Behavioral Sciences* 13: 337–51.

Schopenhauer, A. 1969/1819. *The World as Will and Representation*, Vol. 1. New York: Dover.

Schreber, D. P. 1955/1903. *Memoirs of My Nervous Illness.* London: William Dawson and Sons.

Scull, A. 2015. *Madness in Civilization.* Princeton, NJ: Princeton University Press.

Shorter, E. 2005. *A Historical Dictionary of Psychiatry.* Oxford: Oxford University Press.

Shorter, E. 1997. *A History of Psychiatry.* New York: John Wiley and Sons.

Sinnott-Armstrong, W. (ed.) 2008. *Moral Psychology*, Volume III, *The Neuroscience of Morality: Emotion, Brain Disorders, and Development.* Cambridge, MA: MIT Press.

Smith, D. 1969. Speed freaks vs. acid heads: Conflict between drug subcultures. *Clinical Pediatrics* 8: 185–92.

Snyder, S. H. 1973. Amphetamine psychosis: A "model" schizophrenia mediated by catecholamines. *American Journal of Psychiatry* 130 (1): 61–67.

Snyder, S. H., Banerjee, S. P., Yamamura, H. I., and Greenberg, D. 1974. Drugs, neurotransmitters, and schizophrenia. *Science* 184 (4143): 1243–53.

Sober, E. 1988. Apportioning causal responsibility. *Journal of Philosophy* 85: 303–18.

Spitzer, R. L., and Endicott, J. 1978. Medical and mental disorder: Proposed definition and criteria. In *Critical Issues in Psychiatric Diagnosis*, ed. R. L. Spitzer and D. L. Klein, 15–39. New York: Raven Press.

Spitzer, R. L., Sheehy, M., and Endicott, J. 1977. *DSM-III*: Guiding principles. In *Psychiatric Diagnosis*, ed. V. M. Rakoff, H. C. Stancer, and H. B. Kedward, 1–24. New York: Brunner/Mazel.

Spurzheim, J. G. 1836. *Observations on the Deranged Manifestations of the Mind; or, Insanity*. Boston: Marsh, Capen, and Lyon.

Steinberg, H., and Himmerich, H. 2012. Johann Christian August Heinroth (1773–1843): The first professor of psychiatry as a psychotherapist. *Journal of Religion and Health* 51: 256–68.

Sterelny, K., and Kitcher, P. 1988. The return of the gene. *Journal of Philosophy* 85: 339–61.

Strawson, P. 1962. Freedom and resentment. *Proceedings of the British Academy* 48: 187–221.

Sullivan, H. S. 1962. *Schizophrenia as a Human Process*. New York: W. W. Norton.

Sullivan, H. S. 1956. *Clinical Studies in Psychiatry*. New York: W. W. Norton.

Sullivan, H. S. 1953. *Interpersonal Theory of Psychiatry*. New York: W. W. Norton.

Sullivan, H. S. 1940. *Conceptions of Modern Psychiatry*. New York: W. W. Norton.

Sullivan-Bissett, E. 2018. Monothematic delusion: A case of innocence from experience. *Philosophical Psychology* 31: 920–47.

Tabb, K. 2019. Locke on enthusiasm and the association of ideas. In *Oxford Studies in Early Modern Philosophy*, Vol. IX, ed. D. Rutherford, TK–TK. Oxford: Oxford University Press.

Tabb, K. 2015. Psychiatric progress and the assumption of diagnostic discrimination. *Philosophy of Science* 82: 1047–58.

Taylor, C. 2007. *A Secular Age*. Cambridge, MA: Harvard University Press.

Toohey, P. 2020. *Hold On: The Life, Science, and Art of Waiting*. Oxford: Oxford University Press.

Tsou, J. Y. 2020. Social construction, HPC kinds, and the projectability of human categories. *Philosophy of the Social Sciences* 50: 115–37.

Tsou, J. Y. 2016. *DSM-5* and psychiatry's second revolution: Descriptive vs. theoretical approaches to psychiatric classification. In *The DSM-5 in Perspective*, ed. S. Demazeux and P. Singy, 43–62. Dordrecht: Springer.

Turner, B. S. 1982. The government of the body: Medical regimens and the rationalization of diet. *The British Journal of Sociology* 33: 254–69.

Uhlhaas, P. J., and Mishara, A. L. 2007. Perceptual anomalies in schizophrenia: Integrating phenomenology and cognitive neuroscience. *Schizophrenia Bulletin* 33: 142–56.

Voltaire. 1904/1764. *The Works of Voltaire: A Contemporary Version*, Volume VII. Akron, OH: Werner.

Wakefield, J. C. 1993. Limits of operationalization: A critique of Spitzer and Endicott's (1978) proposed operational criteria for mental disorder. *Journal of Abnormal Psychology* 102: 160–72.

Wakefield, J. C. 1992. The concept of mental disorder: On the boundary between biological facts and social values. *American Psychologist* 47: 373–88.

Walker, D. P. 1981. *Unclean Spirits: Possession and Exorcism in France and England in the Late Sixteenth and Early Seventeenth Centuries.* Philadelphia: University of Pennsylvania Press.

Weiner, D. B. 2000. Betrayal! The 1806 English translation of Pinel's *Traité médico-philosophique sur l'aliénation mentale ou la manie. Gesnerus* 57: 42–50.

Weinfeld, M. 1973. Covenant terminology in the ancient Near East and its influence on the West. *Journal of the American Oriental Society* 92: 190–99.

Wigan, A. L. 1844. *A New View of Insanity.* London: Longman, Brown, Green, and Longmans.

Woolfolk, R. L. 1999. Malfunction and mental disorder. *The Monist* 82: 658–70.

Woolley, D. W., and Shaw, E. 1954. A biochemical and pharmacological suggestion about certain mental disorders. *Science* 119 (3096): 577–78.

Wright, L. 1976. *Teleological Explanations.* Berkeley: University of California Press.

Index